AF675170

Geochemistry and Environmental Geology

Geochemistry and Environmental Geology

Umesh Kumar

AMIGA PRESS. INC
Delhi -110051

AMIGA
PRESS INC.

AMIGA PRESS INC.
Registered Office:
19 DBS Building, Ansari Road, Darya Ganj, Delhi. 110002, India
Phone: +91-11-23283333 Email: amigapress@gmail.com

This edition is published by Amiga Press.Inc

ISBN : 978-93-84533-10-6

Copyright© Reserved
First Published 2018

Reprint 2026

Publisher's note:
This publication may not reproduced, stored in a retrieval system or transmitted in any form or by any means, photocopying , mechanical, electronic, recording or otherwise, without the prior permission from publication house.

Printed and bound in India by Replika Press Pvt. Ltd.

Preface

Geochemistry is the science that uses the tools and principles of chemistry to explain the mechanisms behind major geological systems such as the Earth's crust and its oceans. Multidisciplinary approaches are used to study the interactions between humans and their environments, including the biosphere, lithosphere, hydrosphere and the atmosphere. Research topics include contaminant transport, management of geologic and hydrologic resources, assessment of the risk of exposure to potential hazards and the potential for mitigative actions.

Geochemically scarce metals rarely form minerals in common rocks. Instead, they are carried in the structures of common rock-forming minerals (most of them silicates) through the process of atomic substitution. This process involves the random replacement of an atom in a mineral by a foreign atom of similar ionic radius and valence, without changing the atomic packing of the host mineral. Atoms of copper, zinc, and nickel, for example, can substitute for iron and magnesium atoms in olivine and pyroxene. However, since substitution of foreign atoms produces strains in an atomic packing, there are limits to this process, as determined by temperature, pressure, and various chemical parameters. One important consequence that derives from the way abundant and scarce metals occur in common rocks is that ore minerals of abundant metals can be found in many common rocks, while ore minerals of scarce metals can be found only where some special, restricted geologic process has formed localized enrichments that exceed the limits of atomic substitution.

Environmental geology, like hydrogeology, is an applied science concerned with the practical application of the principles of geology in the solving of environmental problems. It is a multidisciplinary field that is closely related to engineering geology and, to a lesser extent, to environmental geography.

Engineering geologic studies may be performed during the planning, environmental impact analysis, civil or structural engineering design, value engineering and construction phases of public and private works projects, and during post-construction and forensic phases of projects. Works completed by engineering geologists include; geologic hazards, geotechnical, material properties, landslide and slope stability, erosion, flooding, dewatering, andseismic investigations, etc. Engineering geologic studies are performed by a geologist or engineering geologist that is educated, trained and has obtained experience related to the recognition and interpretation of natural processes, the understanding of how these processes impact man-made structures (and vice versa), and knowledge of methods by which to mitigate for hazards resulting from adverse natural or man-made conditions. The principal objective of the engineering geologist is the protection of life and property against damage caused by geologic conditions.

Engineering geologic practice is also closely related to the practice of geological engineering, geotechnical engineering, soils engineering, environmental geology and economic geology. If there is a difference in the content of the disciplines described, it mainly lies in the training or experience of the practitioner.

This book is a compilation of available information, output of research and field experiences of the authors on the subject, that includes techniques to protect the ecosystem and hence to prevent land degradation.

—Editor

Contents

1

Geochemistry

Geochemistry is the science that uses the tools and principles of chemistry to explain the mechanisms behind major geological systems such as the Earth's crust and its oceans. The realm of geochemistry extends beyond the Earth, encompassing the entire Solar System and has made important contributions to the understanding of a number of processes including mantle convection, the formation of planets and the origins of granite and basalt.

GEOCHEMICALLY ABUNDANT AND SCARCE METALS

Metals used in industrial and technological applications can be divided into two classes on the basis of their abundance in the Earth's crust. The geochemically abundant metals, of which there are five (aluminum, iron, magnesium, manganese, and titanium), constitute more than 0.1 percent by weight of the Earth's crust, while the geochemically scarce metals, which embrace all other metals (including such familiar ones as copper, lead, zinc, gold, and silver), constitute less than 0.1 percent. In almost every rock, at least tiny amounts of all metals can be detected by sensitive chemical analysis.

However, there are important differences in the way the abundant and scarce metals occur in common rocks. Geochemically abundant metals tend to be present as essential constituents in minerals. For example, basalt, a common igneous rock, consists largely of the minerals olivine and pyroxene (both magnesium-iron silicates), feldspar (sodium-calcium-aluminum silicate), and ilmenite (iron-titanium oxide). Careful chemical analysis of a basalt will reveal the presence of most of the geochemically scarce metals too, but no amount of searching will reveal minerals in which one or more of the scarce metals is an essential constituent.

Geochemically scarce metals rarely form minerals in common rocks. Instead, they are carried in the structures of common rock-forming minerals (most of them silicates) through the process of atomic substitution. This process involves the random replacement of an atom in a mineral by a foreign atom of similar ionic radius and valence, without changing the atomic packing of the host mineral. Atoms of copper, zinc, and nickel, for example, can substitute for iron and magnesium atoms in olivine and pyroxene. However, since substitution of foreign atoms produces strains in an atomic packing, there are limits to this process, as determined by temperature, pressure, and various chemical parameters. One important consequence that derives from the way abundant and scarce metals occur in common rocks is that ore minerals of abundant metals can be found in many common rocks, while ore minerals of scarce metals can be found only where some special, restricted geologic process has formed localized enrichments that exceed the limits of atomic substitution.

Ore Minerals

Two factors determine whether a given mineral is suitable to be an ore mineral. The first is the ease with which a mineral can be separated from the gangue and concentrated for smelting. Concentrating processes, which are based on the physical properties of the mineral, include magnetic separation, gravity separation, and flotation. The second factor is smelting—that is, releasing the metal from the other elements to which it is chemically bonded in the mineral. Smelting processes, but of primary importance in this consideration of the suitability of an ore mineral is the amount of energy needed to break the chemical bonds and release the metal. In general, less energy is needed to smelt sulfide, oxide, or hydroxide minerals than is required to smelt a silicate mineral. For this reason, few silicate minerals are ore minerals. Because the great bulk of the Earth's crust (about 95 percent) is composed of silicate minerals, sulfide, oxide, and hydroxide ore minerals are at best only minor constituents of the Earth's crust—and in many cases they are very rare constituents.

The preferred ore minerals of both geochemically abundant and geochemically scarce metals are native metals, sulfides, oxides, hydroxides, or carbonates. In a few cases, silicate minerals have to be used as ore minerals because the metals either do not form more desirable minerals or form desirable minerals that rarely occur in large deposits.

Native Metals

Only two metals, gold and platinum, are found principally in their native state, and in both cases the native metals are the primary ore minerals. Silver, copper, iron, osmium, and several other metals also occur in the native state, and a few occurrences are large enough—and sufficiently rich—to be ore deposits. One example is the rich deposits of native copper in the Lake Superior area of Michigan in the United States.

Sulfides

The largest group of ore minerals consists of sulfides. Because the physical and chemical properties of telluride, selenide, and arsenide minerals are very similar to those of sulfide minerals, and because these minerals tend to occur together, it is convenient to use the term sulfide to embrace these similar minerals. Copper, lead, zinc, nickel, molybdenum, silver, arsenic, antimony, bismuth, cobalt, and mercury all form sulfide ore minerals. Gold and silver form tellurides under certain circumstances, and platinum forms an important arsenide ore mineral.

The principle of atomic substitution operates in all classes of minerals, and some of the rarest metals occur by atomic substitution in sulfide ore minerals of other scarce metals. For example, cadmium and indium are generally present in small amounts in the zinc sulfide sphalerite, the major ore mineral of zinc. In fact, most of the world's cadmium and indium is recovered as a by-product of the smelting of sphalerite concentrates to produce zinc.

Oxides and Hydroxides

Oxides and hydroxides are a large and diverse group of ore minerals. The major ore minerals of the geochemically abundant metals aluminum, iron, manganese, and titanium are either oxides or hydroxides, while the oxide-forming scarce metals are chromium, tin, tungsten, tantalum, niobium, and uranium. Vanadium is found mainly by atomic substitution in magnetite, a major oxide ore mineral of iron.

CHEMICAL CHARACTERISTICS

The more common rock constituents are nearly all oxides; chlorides, sulfides and fluorides are the only important exceptions to this and their total amount in any rock is usually much less than 1%. F. W. Clarke has calculated that a little more than 47% of the Earth's crust consists of oxygen. It occurs principally in combination as oxides, of which the chief are silica, alumina, iron oxides, and various carbonates (calcium carbonate, magnesium carbonate, sodium carbonate, and potassium carbonate). The silica functions principally as an acid, forming silicates, and all the commonest minerals of igneous rocks are of this nature. From a computation based on 1672 analyses of numerous kinds of rocks Clarke arrived at the following as the average percentage composition: SiO_2=59.71, Al_2O_3=15.41, Fe_2O_3=2.63, FeO=3.52, MgO=4.36, CaO=4.90, Na_2O=3.55, K_2O=2.80, H_2O=1.52, TiO_2=0.60, P_2O_5=0.22, total 99.22%). All the other constituents occur only in very small quantities, usually much less than 1%.

These oxides combine in a haphazard way. For example, potash (potassium carbonate) and soda (sodium carbonate) combine to produce feldspars. In some cases they may take other forms, such as nepheline, leucite, and muscovite, but in the great majority of instances they are found as feldspar.Phosphoric acid with lime (calcium carbonate) forms apatite. Titanium dioxide with ferrous oxide gives rise to ilmenite. Part of the lime forms lime feldspar. Magnesium carbonate and iron oxides with silica crystallize as olivine or enstatite, or with alumina and lime form the complex ferro-magnesian silicates of which the pyroxenes, amphiboles, and biotites are the chief. Any excess of silica above what is required to neutralize the bases will separate out as quartz; excess of alumina crystallizes as corundum. These must be regarded only as general tendencies. It is possible, by rock analysis, to say approximately what minerals the rock contains, but there are numerous exceptions to any rule.

Mineral Constitution

Hence we may say that except in acid or siliceous rocks containing 66% of silica and over, quartz will not be abundant. In basic rocks (containing 20% of silica or less) it is rare and accidental. If magnesia and iron be above the average while silica is low, olivine may be expected; where silica is present in greater quantity over ferro-magnesian minerals, such as augite, hornblende, enstatite or biotite, occur rather than olivine. Unless potash is high and silica relatively low, leucite will not be present, for leucite does not occur with free quartz. Nepheline, likewise, is usually found in rocks with much soda and comparatively little silica.

With high alkalis, soda-bearing pyroxenes and amphiboles may be present. The lower the percentage of silica and the alkalis, the greater is the prevalence of calcium feldspar as contracted with soda or potash feldspar. Clarke has calculated the relative abundance of the principal rock-forming minerals with the following results: apatite=0.6, titanium minerals=1.5, quartz=12.0, feldspars=59.5, biotite=3.8, hornblende and pyroxene=16.8, total=94.2%. This, however, can only be a rough approximation.

The other determining factor, namely the physical conditions attending consolidation, plays on the whole a smaller part, yet is by no means negligible, as a few instances will prove. Certain minerals are practically confined to deep-seated intrusive rocks, e.g., microcline, muscovite, diallage. Leucite is very rare in plutonic masses; many minerals have special peculiarities in microscopic character according to whether they crystallized in depth or near the surface, e.g., hypersthene, orthoclase, quartz. There are some curious instances of rocks having the same chemical composition, but consisting of entirely different minerals, e.g., the hornblendite of Gran, in Norway, which contains only hornblende, has the same composition as some of the camptonites of the same locality that contain feldspar and hornblende of a different variety. In this connection we may repeat what has been said above about the corrosion of porphyritic minerals in igneous rocks.

In rhyolites and trachytes, early crystals of hornblende and biotite may be found in great numbers partially converted into augite and magnetite. Hornblende and biotite were stable under the pressures and other conditions below the surface, but unstable at higher levels. In the ground-mass of these rocks, augite is almost universally present. But the plutonic representatives of the same magma, granite and syenite contain biotite and hornblende far more commonly than augite.

Felsic, Intermediate and Mafic Igneous Rocks

Those rocks that contain the most silica, and on crystallizing yield free quartz, form a group generally designated the "felsic" rocks. Those again that contain least silica and most magnesia and iron, so that quartz is absent while olivine is usually abundant, form the "mafic" group. The "intermediate" rocks include those characterized by the general absence of both quartz and olivine. An important subdivision of these contains a very high percentage of alkalis, especially soda, and consequently has minerals such as nepheline and leucite not common in other rocks. It is often separated from the others as the "alkali" or "soda" rocks, and there is a corresponding series of mafic rocks. Lastly a small sub-group rich in olivine and without feldspar has been called the "ultramafic" rocks. They have very low percentages of silica but much iron and magnesia.

Except these last, practically all rocks contain felspars or feldspathoid minerals. In the acid rocks the common feldspars are orthoclase, perthite, microcline, and oligoclase—all having much silica and alkalis. In the mafic rocks labradorite, anorthite and bytownite prevail, being rich in lime and poor in silica, potash and soda. Augite is the most common ferro-magnesian in mafic rocks, but biotite and hornblende are on the whole more frequent in felsic rocks.

Rocks that contain leucite or nepheline, either partly or a wholly replacing felspar, are not included in this table. They are essentially of intermediate or of mafic character. We might in consequence regard them as varieties of syenite, diorite, gabbro, etc., in which feldspathoid minerals occur, and indeed there are many transitions between syenites of ordinary type and nepheline – or leucite – syenite, and between gabbro or dolerite and theralite or essexite. But, as many minerals develop in these "alkali" rocks that are uncommon elsewhere, it is convenient in a purely formal classification like that outlined here to treat the whole assemblage as a distinct series.

This classification is based essentially on the mineralogical constitution of the igneous rocks. Any chemical distinctions between the different groups, though implied, are relegated to a subordinate position. It is admittedly artificial but it has grown up with the growth of the science and is still adopted as the basis on which more minute subdivisions are erected. The subdivisions are by no means of equal value.

The syenites, for example, and the peridotites, are far less important than the granites, diorites and gabbros. Moreover, the effusive andesites do not always correspond to the plutonic diorites but partly also to the gabbros. As the different kinds of rock, regarded as aggregates of minerals, pass gradually into one another, transitional types are very common and are often so important as to receive special names. The quartz-syenites and nordmarkites may be interposed between granite and syenite, the tonalites and adamellites between granite and diorite, the monzoaites between syenite and diorite, norites and hyperites between diorite and gabbro, and so on.

2

Fundamentals of Stable Isotope Geochemistry

Isotopes are atoms of the same element that have the same numbers of protons and electrons but different numbers of neutrons. The difference in the number of neutrons between the various isotopes of an element means that the various isotopes have similar charges but different masses. The superscript number to the left of the element designation indicates the number of protons plus neutrons in the isotope.

For example, among the hydrogen isotopes, deuterium (denoted as D or ^{2}H) has one neutron and one proton. This is approximately twice the mass of protium (^{1}H) whereas tritium (^{3}H) has two neutrons and is approximately three times the mass of protium. All isotopes of oxygen have 8 electrons and 8 protons; however, an oxygen atom with a mass of 18 (denoted ^{18}O) has 2 more neutrons than oxygen-16 (^{16}O).

The original isotopic compositions of planetary systems are a function of nuclear processes in stars. Over time, isotopic compositions in terrestrial environments change by the processes of radioactive decay, cosmic ray interactions, and such anthropogenic activities as processing of nuclear fuels, reactor accidents, and nuclear-weapons testing. Radioactive (unstable) isotopes are nuclei that spontaneously disintegrate over time to form other isotopes. During the disintegration, radioactive isotopes emit alpha or beta particles and sometimes also gamma rays.

The so-called stable isotopes are nuclei that do not appear to decay to other isotopes on geologic timescales, but may themselves be produced by the decay of radioactive isotopes. For example, ^{14}C, a radioisotope of carbon, is produced in the atmosphere by the interaction cosmic-ray neutrons with stable ^{14}N. With a half-life of about 5730 years, ^{14}C decays back to ^{14}N by emission of a beta particle; the stable ^{14}N produced by radioactive decay is called "radiogenic" nitrogen.

Isotope Terminology

In everyday speech, isotopes are still described with the element name given first, as in "oxygen-18" or "O-18" instead of "18- oxygen". And many texts, especially older ones that were typeset without superscripts, show the mass number to the right of the element abbreviation, as in C-13 or C13 for carbon-13. However, both in speech and in media, it is becoming more common to put the mass number before the element name, as is ^{15}N.

The stable isotopic compositions of low-mass (light) elements such as oxygen, hydrogen, carbon, nitrogen, and sulfur are normally reported as "delta" (d) values in parts per thousand (denoted as ‰) enrichments or depletions relative to a standard of known composition. The symbol ‰ is spelled out in several different ways: permil, per mil, per mill, or per mille. The term "per mill" is the ISO term, but is not yet widely used. d values are calculated by:

$$\text{(in ‰)} = (R_{sample}/R_{standard} - 1)1000$$

where "R" is the ratio of the heavy to light isotope in the sample or standard. For the elements sulfur, carbon, nitrogen, and oxygen, the average terrestrial abundance ratio of the heavy to the light isotope ranges from 1:22 (sulfur) to 1:500 (oxygen); the ratio ^{2}H:^{1}H is 1:6410. A positive d value means that the sample contains more of the heavy isotope than the standard; a negative dvalue means that the sample contains less of the heavy isotope than the standard. A d^{15}N value of +30‰ means that there are 30 parts-per-thousand or 3% more ^{15}N in the sample relative to the standard.

In ASCII-only media, the term delta is almost always denoted with the small Greek letter d. In media lacking this symbol, it is not- uncommonly replaced informally with the letter "d". The term d is spelled and pronounced delta not del. The word del describes either of two mathematical terms: an operator or a partial derivative.

Many isotopers are very sensitive about misuses of isotope terminology.Harmon Craig's immortal limerick says it all:

There was was a young man from Cornell
Who pronounced every "delta" as "del"
But the spirit of Urey
Returned in a fury
And transferred that fellow to hell

There are several commonly used ways for making comparisons between the dvalues of two materials. The first three are preferred.

- higher vs. lower values
- heavier vs. lighter (the "heavier" material is the one with the higher value)

- more/less positive vs. more/less negative (eg., -10‰ is more positive than -20‰)
- enriched vs. depleted (remember to state what isotope is in short supply; eg., a material is enriched in ^{18}O or ^{16}O relative to some other material).

Basic Principles

The various isotopes of an element have slightly different chemical and physical properties because of their mass differences. For elements of low atomic numbers, these mass differences are large enough for many physical, chemical, and biological processes or reactions to "fractionate" or change the relative proportions of various isotopes. Two different types of processes — equilibrium and kinetic isotope effects — cause isotope fractionation. As a consequence of fractionation processes, waters and solutes often develop unique isotopic compositions (ratios of heavy to light isotopes) that may be indicative of their source or of the processes that formed them.

Equilibrium isotope-exchange reactions involve the redistribution of isotopes of an element among various species or compounds. At equilibrium, the forward and backward reaction rates of any particular isotope are identical. This does not mean that the isotopic compositions of two compounds at equilibrium are identical, but only that the ratios of the different isotopes in each compound are constant. During equilibrium reactions, the heavier isotope generally becomes enriched (preferentially accumulates) in the species or compound with the higher energy state. For example, sulfate is enriched in ^{34}S relative to sulfide; consequently, the sulfide is described as depleted in ^{34}S relative to sulfate. During phase changes, the ratio of heavy to light isotopes in the molecules in the two phases changes. For example, as water vapor condenses (an equilibrium process), the heavier water isotopes (^{18}O and ^{2}H) become enriched in the liquid phase while the lighter isotopes (^{16}O and ^{1}H) tend toward the vapor phase.

Kinetic isotope fractionations occur in systems out of isotopic equilibrium where forward and backward reaction rates are not identical. The reactions may, in fact, be unidirectional if the reaction products become physically isolated from the reactants. Reaction rates depend on the ratios of the masses of the isotopes and their vibrational energies; as a general rule, bonds between the lighter isotopes are broken more easily than the stronger bonds between the heavy isotopes. Hence, the lighter isotopes react more readily and become concentrated in the products, and the residual reactants become enriched in the heavy isotopes.

Biological processes are generally unidirectional and are excellent examples of "kinetic" isotope reactions. Organisms preferentially use the lighter isotopic species because of the lower energy "costs", resulting in significant fractionations between the substrate (heavier) and the biologically mediated product (lighter). The magnitude of the fractionation depends on the reaction pathway utilized and the relative energies of the bonds being severed and formed by the reaction. In general, slower reaction steps show greater isotopic fractionation than faster steps because the organism has time to be more selective. Kinetic reactions can result in fractionations very different from, and typically larger than, the equivalent equilibrium reaction.

Many reactions can take place either under purely equilibrium conditions or be affected by an additional kinetic isotope fractionation. For example, although evaporation can take place under purely equilibrium conditions (i.e., at 100% humidity when the air is still), more typically the products become partially isolated from the reactants (e.g., the resultant vapor is blown downwind). Under these conditions, the isotopic compositions of the water and vapor are affected by an additional kinetic isotope fractionation of variable magnitude.

The partitioning of stable isotopes between two substances A and B can be expressed by use of the isotopic fractionation factor (alpha):

$$A\text{-}B = R_A/R_B$$

where "R" is the ratio of the heavy to light isotope (e.g., $^2H/^1H$ or $^{18}O/^{16}O$). Values for alpha tend to be very close to 1. Kinetic fractionation factors are typically described in terms of enrichment or discrimination factors; these are defined in various ways by different researchers.

Isotopic compositions are determined in specialized laboratories using isotope ratio mass spectrometry. The analytical precisions are small relative to the ranges in d values that occur in natural earth systems.

Typical one standard deviation analytical precisions for oxygen, carbon, nitrogen, and sulfur isotopes are in the range of 0.05‰ to 0.2‰; typical precisions for hydrogen isotopes are poorer, from 0.2 to 2.0‰, because of the lower ^2H:^1H ratio.

Stable Isotope Standards

Various isotope standards are used for reporting isotopic compositions; the compositions of each of the standards have been defined as 0‰. Stable oxygen and hydrogen isotopic ratios are normally reported relative to the SMOW standard ("Standard Mean Ocean Water" or the virtually equivalent VSMOW (Vienna-SMOW) standard.

Carbon stable isotope ratios are reported relative to the PDB (for Pee Dee Belemnite) or the equivalent VPDB (Vienna PDB) standard. The oxygen stable isotope ratios of carbonates are commonly reported relative to PDB or VPDB, also. Sulfur and nitrogen isotopes are reported relative to CDT (for Cañon Diablo Troilite) and AIR (for atmospheric air), respectively.

VSMOW and VPDB are virtually identical to the now-unavailable SMOW and PDB standards. However, use of VSMOW and VPDB is preferred (and in some journals is now required) because their use implies that the measurements have been calibrated according to IAEA (International Atomic Energy Agency) guidelines for expression of delta values relative to available reference materials on normalized permil scales (Coplen, 1996). Laboratories accustomed to analyzing anthropogenic, highly enriched compounds may report absolute isotope abundances in atomic-weight percent or ppm instead of ratios in per mil.

FUNDAMENTALS OF ISOTOPE GEOCHEMISTRY

The fundamentals of stable and radio-isotope geochemistry, intended to provide readers with the necessary background information to understand the succeeding chapters.

Definitions

Isotopes are atoms of the same element that have different numbers of neutrons. Differences in the number of *neutrons* among the various isotopes of an element mean that the various isotopes have different masses. The superscript number to the left of the element designation is called the *mass number* and is the sum of the number of protons and neutrons in the isotope. For example, among the hydrogen isotopes, deuterium (denoted as D or ^{2}H) has one neutron and one proton. This mass number of 2 is approximately equal to twice the mass of protium (^{1}H), whereas tritium (^{3}H) has two neutrons and its mass is approximately three times the mass of protium. Isotopes of the same element have the same number of protons. For example, all isotopes of oxygen have 8 protons; however, an oxygen atom with a mass of 18 (denoted ^{18}O) has 2 more neutrons than oxygen with a mass of 16 (^{16}O). Isotope names are usually pronounced with the element name first, as in "oxygen-18" instead of "18-oxygen." In many texts, especially older ones typeset without superscripts, the mass number is shown to the right of the element abbreviation, as in C-13 or C^{13} for carbon-13.

The original isotopic compositions of planetary systems are a function of nuclear processes in stars. Over time, isotopic compositions in terrestrial environments change by the processes of radioactive decay, cosmic ray interactions, mass-dependent fractionations that accompany inorganic and biological reactions, and anthropogenic activities such as the processing of nuclear fuels, reactor accidents, and nuclear-weapons testing. *Radioactive* (unstable) isotopes are *nuclides* (isotope-specific atoms) that spontaneously disintegrate over time to form other isotopes. During the disintegration, radioactive isotopes emit *alpha* or *beta* particles and sometimes also *gamma rays*. *Stable* isotopes are nuclides that do not appear to decay to other isotopes on geologic time scales, but may themselves be produced by the decay of radioactive isotopes.

Naturally occurring nuclides define a path in the chart of nuclides, corresponding to the greatest stability of the neutron/proton (N/Z) ratio. For nuclides of low atomic mass, the greatest stability is achieved when the number of neutrons and protons are approximately equal ($N = Z$); these are the so-called *stable* isotopes. However, as the atomic mass increases, the stable neutron/proton ratio increases until $N/Z = 1.5$. Radioactive decay occurs when changes in N and Z of an unstable nuclide cause the trans-formation of an atom of one nuclide into that of another, more stable nuclide; these radioactive nuclides are called *unstable* nuclides.

Atoms produced by the radioactive decay of other nuclides are termed *radiogenic*. A few nuclides are produced by cosmic ray bombardment of stable nuclides in the atmosphere and are termed *cosmogenic*. Other nuclides may be created by the addition of neutrons produced by the alpha decay of other nuclides (neutron activation).

Alternatively, the neutron addition can displace a proton in the nucleus, creating a nuclide of the same atomic mass but lower atomic number. Nuclides produced by these two processes are termed *lithogenic*. If the daughter product is radioactive, it will decay to form an isotope of yet another element. This process will continue until a stable nuclide is produced. For example, uranium and thorium decay to form other radionuclides that are themselves radioactive and decay to other radionuclides, and so on until stable lead isotopes are formed. Although the terms *parent* and *daughter* nuclides are commonly used, these terms can be misleading. Only *one* atom is involved during radioactive decay; that is, the daughter nuclide is the same nuclide as the parent atom. However, after radioactive decay it has a different number of neutrons in its nucleus.

The change in the number of neutrons can occur in a variety of ways. However, the four mechanisms described below are the most common and produce the radiogenic nuclides most relevant to hydrologic and geologic studies:

Beta decay occurs when nuclides deficient in protons transform a neutron into a proton and an electron, and expel the electron from the nucleus as a negative b particle (b^-), thereby increasing the atomic number by one while the number of neutrons is reduced by one.

Positron decay occurs when nuclides deficient in neutrons transform a proton into a neutron, an electron (b^+), and a neutrino, thereby decreasing the atomic number by one and increasing the neutron number by one. The daughters are *isobars* (nuclides of equal mass) of their parent and are isotopes of different elements.

Beta capture (or electron capture) occurs when nuclides deficient in neutrons transform a proton into a neutron plus neutrino by the capture of an electron by a proton, thereby decreasing the number of protons in the nucleus by one. Both this and positron decay yield a radiogenic nuclide that is an isobar of the parent nuclide.

Alpha decay occurs when heavy atoms above Z = 83 in the nuclide chart emit an alpha particle, which consists of a helium nuclei with two neutrons, two protons, and a 2^+ charge.

This radiogenic daughter product in not an isobar of its parent nuclide because its mass is reduced by four.

For example, the radioisotope (radioactive isotope) ^{14}C is produced in the atmosphere by cosmic ray neutron interaction with ^{14}N.

^{14}C has a half-life of about 5730 years, and decays back to stable ^{14}N by emission of a beta particle. The *decay equation* below expresses the change in the concentration (activity) of the nuclide over time:

$$A_t = A_o \cdot e^{-lt}$$

where A_o is the initial activity of the parent nuclide, and A_t is its activity after some time "t." The decay constant "l" is equal to ln ($2/t^{½}$). Note that the decay rate is only a function of the activity of the nuclide and time, and that temperature and other environmental parameters appear to have no effect on the rate.

Terminology

Stable isotope compositions of low-mass (light) elements such as oxygen, hydrogen, carbon, nitrogen, and sulfur are normally reported as d values. The term "d" is spelled and pronounced delta not del. The word del describes either of two mathematical terms: an operator ( ) or a partial derivative (*d*). d values are reported in units of parts per thousand (denoted as ‰ or permil, or per mil, or per mille — or even recently, per mill) relative to a standard of known composition. d values are calculated by:

$$d \text{ (in ‰)} = (R_x / R_s - 1) \cdot 1000$$

where *R* denotes the ratio of the heavy to light isotope (e.g., $^{34}S/^{32}S$), and R_x and R_s are the ratios in the sample and standard, respectively. For sulfur, carbon, nitrogen, and oxygen, the average terrestrial abundance ratio of the heavy to the light isotope ranges from 1:22 (sulfur) to 1:500 (oxygen); the ratio ^2H:^1H is much lower at 1:6410. A positive d value means that the isotopic ratio of the sample is higher than that of the standard; a negative d value means that the isotopic ratio of the sample is lower than that of the standard. For example, a $d^{15}N$ value of +30‰ means that the $^{15}N/^{14}N$ of the sample is 30 parts-per-thousand or 3% higher than the $^{15}N/^{14}N$ of the standard. Many isotope geochemists advocate always prefacing the d value with a sign, even when the value is positive, to distinguish between a true positive d value and a d value that is merely missing its sign (a frequent occurrence with users unfamiliar with isotope terminology).

There are several commonly used ways for making comparisons between the d values of two materials. The first two are preferred because of their clarity, and the fourth should be avoided:

(1) high vs. low values
(2) more/less positive vs. more/less negative (e.g., -10‰ is more positive than -20‰)
(3) heavier vs. lighter (the "heavy" material is the one with the higher d value)
(4) enriched vs. depleted (always remember to state what isotope is in short supply, e.g., a material is enriched in ^{18}O or ^{16}O relative to some other material, *and* that the enrichment or depletion is a result of some reaction or process). For example, to say that "one sample is enriched in ^{34}S relative to another because of sulfate reduction" is proper usage. Phrases such as "a sample has an enriched $d^{15}N$ value" are misuses of terminology.

Standards

The isotopic compositions of materials analyzed on mass spectrometers are usually reported relative to some international *reference standard*. Samples are either analyzed at the same time as this reference standard or with some internal laboratory standard that has been calibrated relative to the international standard. Alternatively, the absolute ratios of isotopes can be reported. Small quantities of these reference standards are available for calibration purposes from either the National Institute of Standards and Technology (NIST) in the USA, or the International Atomic Energy Agency (IAEA) in Vienna.

Various isotope standards are used for reporting light stable-isotopic compositions. The d values of each of the standards have been defined as 0‰. dD and $d^{18}O$ values are normally reported relative to the SMOW standard (Standard Mean Ocean Water; Craig, 1961) or the equivalent VSMOW (Vienna-SMOW) standard. $d^{13}C$ values are reported relative to either the PDB (Pee Dee Belemnite) or the equivalent VPDB (Vienna-PDB) standard. $d^{18}O$ values of low-temperature carbonates are also commonly reported relative to PDB or VPDB.

VSMOW and VPDB are virtually identical to the SMOW and PDB standards. Use of VSMOW and VPDB is supposed to imply that the measurements were calibrated according to IAEA guidelines for expression of dvalues relative to available reference materials on *normalized* permil scales . Laboratories accustomed to analyzing synthetic compounds that are highly enriched in the heavy (or, less commonly, the light) isotope may report absolute isotope abundances in atomic-weight percent or ppm, instead of relative ratios in permil. In general, radioisotopes are reported as absolute concentrations or ratios. Tritium (^{3}H) values are typically reported as absolute concentrations, called Tritium Units (TU) where one TU corresponds to 1 tritium atom per 10^{18} hydrogen atoms. Tritium values may also be expressed in terms of activity (pico-Curies/liter, pCi/L) or decay (disintegrations per minute/liter, dpm/L), where 1 TU = 3.2 pCi/L = 7.2 dpm/L. ^{14}C contents are referenced to an international standard known as "modern carbon" and are typically expressed as a percent of modern carbon (pmc).

RADIOGENIC ISOTOPES

Lead Isotopes

Lead consists of four stable isotopes: ^{204}Pb, ^{206}Pb, ^{207}Pb, and ^{208}Pb. Local variations in uranium/thorium/lead content cause a wide location-specific variation of isotopic ratios for lead from different localities. Lead emitted to the atmosphere by industrial processes has an isotopic composition different from lead in minerals. Combustion of gasoline with tetraethyllead additive led to formation of ubiquitous micrometer-sized lead-rich particulates in car exhaust smoke; especially in urban areas the man-made lead particles are much more common than natural ones. The differences in isotopic content in particles found in objects can be used for approximate geolocation of the object's origin.

RADIOACTIVE ISOTOPES

Hot particles, radioactive particles of nuclear fallout and radioactive waste, also exhibit distinct isotopic signatures. Their radionuclide composition (and thus their age and origin) can be determined by mass spectrometry or by gamma spectrometry. For example, particles generated by a nuclear blast contain detectable amounts of ^{60}Co and ^{152}Eu. The Chernobyl accident did not release these particles but did release ^{125}Sb and ^{144}Ce. Particles from underwater bursts will consist mostly of irradiated sea salts. Ratios of $^{152}Eu/^{155}Eu$, $^{154}Eu/^{155}Eu$, and $^{238}Pu/^{239}Pu$ are also different for fusion and fissionnuclear weapons, which allows identification of hot particles of unknown origin.

APPLICATIONS

Forensics

With the advent of stable isotope ratio mass spectrometry, isotopic signatures of materials find increasing use in forensics, distinguishing the origin of otherwise similar materials and tracking the materials to their common source. For example the isotope signatures of plants can be to a degree influenced by the growth conditions, including moisture and nutrient availability.

In case of synthetic materials, the signature is influenced by the conditions during the chemical reaction. The isotopic signature profiling is useful in cases where other kinds of profiling, e.g. characterization of impurities, are not optimal. Electronics coupled with scintillator detectors are routinely used to evaluate isotope signatures and identify unknown sources. (one example - SAM Isotope Identifier)

A study was published demonstrating the possibility of determination of the origin of a common brown PSA packaging tape by using the carbon, oxygen, and hydrogen isotopic signature of the backing polymer, additives, and adhesive.

Measurement of carbon isotopic ratios can be used for detection of adulteration of honey. Addition of sugars originated from corn or sugar cane (C4 plants) skews the isotopic ratio of sugars present in honey, but does not influence the isotopic ratio of proteins; in an unadulterated honey the carbon isotopic ratios of sugars and proteins should match. As low as 7% level of addition can be detected.

Nuclear explosions form ^{10}Be by a reaction of fast neutrons with ^{13}C in the carbon dioxide in air. This is one of the historical indicators of past activity at nuclear test sites.

Solar System Origins

Isotopic fingerprints are used to study the origin of materials in the Solar System. For example, the Moon's oxygen isotopic ratios seem to be essentially identical to Earth's. Oxygen isotopic ratios, which may be measured very precisely, yield a unique and distinct signature for each solar system body. Different oxygen isotopic signatures can indicated the origin of material ejected into space.

The Moon's titanium isotope ratio ($^{50}Ti/^{47}Ti$) appears close to the Earth's (within 4 ppm). In 2013, a study was released that indicated water in lunar magma was 'indistinguishable' from carbonaceous chondrites and nearly the same as Earth's, based on the composition of water isotopes.

STABLE ISOTOPE GEOCHEMISTRY

For most stable isotopes, the magnitude of fractionation from kinetic and equilibrium fractionation is very small; for this reason, enrichments are typically reported in "per mil" (‰, parts per thousand). These enrichments represent the ratio of heavy isotope to light isotope in the sample over the ratio of a standard.

Carbon

Carbon has two stable isotopes, ^{12}C and ^{13}C, and one radioactive isotope, ^{14}C. Carbon isotope ratios are measured against Vienna Pee Dee Belemnite(VPDB). They have been used to track ocean circulation, among other things. Carbon stable isotopes are fractionated primarily by photosynthesis (Faure, 2004).

The $^{13}C/^{12}C$ ratio is also an indicator of paleoclimate: a change in the ratio in the remains of plants indicates a change in the amount of photosynthetic activity, and thus in how favorable the environment was for the plants.

During photosynthesis, organisms using the C_3 pathway show different enrichments compared to those using the C_4 pathway, allowing scientists not only to distinguish organic matter from abiotic carbon, but also what type of photosynthetic pathway the organic matter was using.

Nitrogen

Nitrogen has two stable isotopes, ^{14}N, and ^{15}N. The ratio between these is measured relative to nitrogen in ambient air. Nitrogen ratios are frequently linked to agricultural activities.

Nitrogen isotope data has also been used to measure the amount of exchange of air between the stratosphere andtroposphere using data from the greenhouse gas N_2O.

Oxygen

Oxygen has three stable isotopes, ^{16}O, ^{17}O, and ^{18}O. Oxygen ratios are measured relative to Vienna Standard Mean Ocean Water (VSMOW) or Vienna Pee Dee Belemnite (VPDB). Variations in oxygen isotope ratios are used to track both water movement, paleoclimate, and atmospheric gases such as ozone and carbon dioxide.

Typically, the VPDB oxygen reference is used for paleoclimate, while VSMOW is used for most other applications. Oxygen isotopes appear in anomalous ratios in atmospheric ozone, resulting from mass-independent fractionation. Isotope ratios in fossilizedforaminifera have been used to deduce the temperature of ancient seas.

Sulfur

Sulfur has four stable isotopes, with the following abundances: ^{32}S (0.9502), ^{33}S (0.0075), ^{34}S (0.0421) and ^{36}S (0.0002). These abundances are compared to those found in Cañon Diablo troilite. Variations in sulfur isotope ratios are used to study the origin of sulfur in an orebody and the temperature of formation of sulfur–bearing minerals.

RADIOGENIC ISOTOPE GEOCHEMISTRY

Radiogenic isotopes provide powerful tracers for studying the ages and origins of Earth systems.

They are particularly useful to understand mixing processes between different components, because (heavy) radiogenic isotope ratios are not usually fractionated by chemical processes.

Radiogenic isotope tracers are most powerful when used together with other tracers: The more tracers used, the more control on mixing processes. An example of this application is to the evolution of the Earth's crust and Earth's mantle through geological time.

Lead-lead Isotope Geochemistry

Lead has four stable isotopes - ^{204}Pb, ^{206}Pb, ^{207}Pb, ^{208}Pb and one common radioactive isotope ^{202}Pb with a half-life of ~53,000 years. Lead is created in the Earth via decay of transuranic elements, primarily uranium and thorium.

Lead isotope geochemistry is useful for providing isotopic dates on a variety of materials. Because the lead isotopes are created by decay of different transuranic elements, the ratios of the four lead isotopes to one another can be very useful in tracking the source of melts in igneous rocks, the source ofsediments and even the origin of people via isotopic fingerprinting of their teeth, skin and bones.

It has been used to date ice cores from the Arctic shelf, and provides information on the source of atmospheric lead pollution.

Lead-lead isotopes has been successfully used in forensic science to fingerprint bullets, because each batch of ammunition has its own peculiar $^{204}Pb/^{206}Pb$ vs $^{207}Pb/^{208}Pb$ ratio.

Samarium-neodymium

Samarium-neodymium is an isotope system which can be utilised to provide a date as well as isotopic fingerprints of geological materials, and various other materials including archaeological finds (pots, ceramics).

^{147}Sm decays to produce ^{143}Nd with a half life of 1.06×10^{11} years.

Dating is achieved usually by trying to produce an isochron of several minerals within a rock specimen. The initial $^{143}Nd/^{144}Nd$ ratio is determined.This initial ratio is modelled relative to CHUR - the Chondritic Uniform Reservoir - which is an approximation of the chondritic material which formed the solar system. CHUR was determined by analysing chondrite and achondrite meteorites.

The difference in the ratio of the sample relative to CHUR can give information on a model age of extraction from the mantle (for which an assumed evolution has been calculated relative to CHUR) and to whether this was extracted from a granitic source (depleted in radiogenic Nd), the mantle, or an enriched source.

Rhenium-osmium

Rhenium and osmium are siderophile elements which are present at very low abundances in the crust. Rhenium undergoes radioactive decay to produce osmium. The ratio of non-radiogenic osmium to radiogenic osmium throughout time varies.

Rhenium prefers to enter sulfides more readily than osmium. Hence, during melting of the mantle, rhenium is stripped out, and prevents the osmium-osmium ratio from changing appreciably. This *locks in* an initial osmium ratio of the sample at the time of the melting event. Osmium-osmium initial ratios are used to determine the source characteristic and age of mantle melting events.

Noble Gas Isotopes

Natural isotopic variations amongst the noble gases result from both radiogenic and nucleogenic production processes. Because of their unique properties, it is useful to distinguish them from the conventional radiogenic isotope systems described above.

Helium-3

Helium-3 was trapped in the planet when it formed. Some ^{3}He is being added by meteoric dust, primarily collecting on the bottom of oceans (although due to subduction, all oceanic tectonic plates are younger than continental plates). However, ^{3}He will be degassed from oceanic sediment during subduction, so cosmogenic ^{3}He is not affecting the concentration or noble gas ratios of the mantle.

Helium-3 is created by cosmic ray bombardment, and by lithium spallation reactions which generally occur in the crust. Lithium spallation is the process by which a high-energy neutron bombards a lithium atom, creating a ^{3}He and a ^{4}He ion. This requires significant lithium to adversely affect the $^{3}He/^{4}He$

ratio. All degassed helium is lost to space eventually, due to the average speed of helium exceeding the escape velocity for the Earth. Thus, it is assumed the helium content and ratios of Earth's atmosphere have remained essentially stable.

It has been observed that ^3He is present in volcano emissions and oceanic ridge samples. How ^3He is stored in the planet is under investigation, but it is associated with the mantle and is used as a marker of material of deep origin.

Due to similarities in helium and carbon in magma chemistry, outgassing of helium requires the loss of volatile components (water, carbon dioxide) from the mantle, which happens at depths of less than 60 km.

However, ^3He is transported to the surface primarily trapped in the crystal lattice of minerals withinfluid inclusions.

Helium-4 is created by radiogenic production (by decay of uranium/thorium-series elements). The continental crust has become enriched with those elements relative to the mantle and thus more He^4 is produced in the crust than in the mantle.

The ratio (R) of ^3He to ^4He is often used to represent ^3He content. R usually is given as a multiple of the present atmospheric ratio (Ra).

Common values for R/Ra:

- Old continental crust: less than 1
- mid-ocean ridge basalt (MORB): 7 to 9
- Spreading ridge rocks: 9.1 plus or minus 3.6
- Hotspot rocks: 5 to 42
- Ocean and terrestrial water: 1
- Sedimentary formation water: less than 1
- Thermal spring water: 3 to 11

$^3He/^4He$ isotope chemistry is being used to date groundwaters, estimate groundwater flow rates, track water pollution, and provide insights intohydrothermal processes, igneous geology and ore genesis.

- (U-Th)/He dating of apatite as a thermal history tool
- USGS: Helium Discharge at Mammoth Mountain Fumarole (MMF)

URANIUM-SERIES ISOTOPES

U-series isotopes are unique amongst radiogenic isotopes because, being in the U-series decay chains, they are both radiogneic and radioactive. Because their abundances are normally quoted as activity ratios rather than atomic ratios, they are best considered separately from the other radiogenic isotope systems.

Protactinium/Thorium - ^{231}Pa / ^{230}Th

Uranium is well mixed in the ocean, and its decay produces ^{231}Pa and ^{230}Th at a constant activity ratio (0.093). The decay products are rapidly removed byadsorption on settling particles, but not at equal rates. ^{231}Pa has a residence equivalent to the residence time of deep water in the Atlantic basin (around 1000 yrs) but ^{230}Th is removed more rapidly (centuries). Thermohaline circulation effectively exports ^{231}Pa from the Atlantic into the Southern Ocean, while most of the ^{230}Th remains in Atlantic sediments. As a result, there is a relationship between $^{231}Pa/^{230}Th$ in Atlantic sediments and the rate of overturning: faster overturning produces lower sediment $^{231}Pa/^{230}Th$ ratio, while slower overturning increases this ratio. The combination of ä^{13}C and $^{231}Pa/^{230}Th$ can therefore provide a more complete insight into past circulation changes.

ANTHROPOGENIC ISOTOPES

Tritium/helium-3

Tritium was released to the atmosphere during atmospheric testing of nuclear bombs. Radioactive decay of tritium produces the noble gas helium-3. Comparing the ratio of tritium to helium-3 ($^3H/^3He$) allows estimation of the age of recent ground waters.

- USGS Tritium/Helium-3 Dating
- Hydrologic Isotope Tracers - Helium

SAMARIUM-NEODYMIUM DATING

Samarium-neodymium dating is useful for determining the age relationships of rocks and meteorites, based on decay of a long-lived samarium (Sm) isotope to a radiogenic neodymium (Nd) isotope. Nd isotope ratios are used to provide information on the source of igneous melts as well as to provide age data. The various reservoirs within the solid earth will have different values of initial $^{143}Nd/^{144}Nd$ ratios, especially with reference to the mantle.

The usefulness of Sm-Nd dating is the fact that these two elements are rare earths. They are thus, theoretically, not particularly susceptible to partitioning during melting of silicate rocks. The fractionation effects of crystallisation of felsic minerals changes the Sm/Nd ratio of the resultant materials. This, in turn, influences the $^{143}Nd/^{144}Nd$ ratios with ingrowth of radiogenic ^{143}Nd.

The mantle is assumed to have undergone chondritic evolution, and thus deviations in initial $^{143}Nd/^{144}Nd$ ratios can provide information as to when a particular rock or reservoir was separated from the mantle within the Earth's past.

In many cases, Sm-Nd and Rb-Sr isotope data are used together.

Sm-Nd radiometric dating

Samarium has five naturally occurring isotopes and neodymium has seven.

The two elements are joined in a parent-daughter relationship by the alpha-decay of ^{147}Sm to ^{143}Nd with a half life of 1.06×10^{11} years.

^{146}Sm is an almost-extinct nuclide which decays via alpha emission to produce ^{142}Nd, with a half-life of 1.08×10^{8} years.

^{146}Sm is itself produced by the decay of ^{150}Gd via alpha-decay with a half-life of 1.79×10^{6} years.

An isochron is calculated normally. As with Rb-Sr and Pb-Pb isotope geochemistry, the initial $^{143}Nd/^{144}Nd$ ratio of the isotope system provides important information on crustal formation and the isotopic evolution of the solar system.

Sm and Nd Geochemistry

The concentration of Sm and Nd in silicate minerals increase with the order in which they crystallise from a magma according to Bowen's reaction series.

Samarium is accommodated more easily into mafic minerals, so a mafic rock which crystallises mafic minerals will concentrate neodymium in the melt phase faster relative to samarium.

Thus, as a rock undergoes fractional crystallization from a mafic to a more felsic composition, the abundance of Sm and Nd changes, as does the ratio between Sm and Nd.

Thus, ultramafic rocks have low Sm and Nd and *high* Sm/Nd ratios. Felsic rocks have high concentrations of Sm and Nd but *low* Sm/Nd ratios (komatiitehas 1.14 parts per million (ppm) Sm and 3.59 ppm Nd versus 4.65 ppm Sm and 21.6 ppm Nd in rhyolite).

The importance of this process is apparent in modeling the age of continental crust formation.

The CHUR Model

Through the analysis of isotopic compositions of neodymium, DePaolo and Wasserburg discovered that terrestrial igneous rocks closely followed theChondritic Uniform Reservoir (CHUR) line.

Chondritic meteorites are thought to represent the earliest (unsorted) material that formed in the solar system before planets formed. They have relatively homogeneous trace element signatures and therefore their isotopic evolution can model the evolution of the whole solar system and of the 'Bulk Earth'. After plotting the ages and initial $^{143}Nd/^{144}Nd$ ratios of terrestrial igneous rocks on a Nd evolution vs. time diagram, DePaolo and Wasserburg determined that Archean rocks had initial Nd isotope ratios very similar to that defined by the CHUR evolution line.

Epsilon Notation

Since $^{143}Nd/^{144}Nd$ departures from the CHUR evolution line are very small, DePaolo and Wasserburg argued that it would be useful to create a form of notation that described $^{143}Nd/^{144}Nd$ in terms of their deviations from the CHUR evolution line. This is called the epsilon notation whereby one epsilon unit represents a one part per 10,000 deviation from the CHUR composition.

Since epsilon units are larger and therefore a more tangible representation of the initial Nd isotope ratio, by using these instead of the initial isotopic ratios, it is easier to comprehend and therefore compare initial ratios of crust with different ages. In addition, epsilon units will normalize the initial ratios to CHUR, thus eliminating any effects caused by various analytical mass fractionation correction methods applied.

Nd Model Ages

Since CHUR defines initial ratios of continental rocks through time, it was deduced that measurements of $^{143}Nd/^{144}Nd$ and $^{147}Sm/^{144}Nd$, with the use of CHUR, could produce model ages for the segregation from the mantle of the melt which formed any crustal rock. This has been termed 't-CHUR'. In order for a T_{CHUR} age to be calculated, fractionation between Nd/Sm would have to have occurred during magma extraction from the mantle to produce a continental rock. This fractionation would then cause a deviation between the crustal and mantle isotopic evolution lines.

The T_{CHUR} age of a rock, can yield a formation age for the crust as a whole if the sample has not suffered disturbance after its formation. Since Sm/Nd are rare-earth elements (REE), their characteristic immobility enables their ratios to resist partitioning during metamorphism and melting of silicate rocks. This therefore allows for crustal formation ages to be calculated, despite any metamorphism the sample has undergone.

ISOTOPIC SIGNATURE

An isotopic signature (also isotopic fingerprint) is a ratio of non-radiogenic 'stable isotopes', stable radiogenic isotopes, or unstable radioactiveisotopes of particular elements in an investigated material. The ratios of isotopes in a sample material are measured by isotope ratio mass spectrometry.

Stable Isotopes

The atomic mass of different isotopes affect their chemical kinetic behavior, leading to natural isotope separation processes.

Similarly, carbon in inorganic carbonates shows little isotopic fractionation, while carbon in materials originated by photosynthesis is depleted of the heavier isotopes. In addition, there are two types of plants with different biochemical pathways; the C3 carbon fixation, where the isotope separation effect is more pronounced, C4 carbon fixation, where the heavier ^{13}C is less depleted, and Crassulacean Acid Metabolism (CAM) plants, where the effect is similar but less pronounced than with C_4 plants. The different isotope ratios for the two kinds of plants propagate through the food chain, thus it is possible to determine if the principal diet of a human or an animal consists primarily of C_3 plants (rice, wheat, soybeans, potatoes) or C_4 plants (corn, or corn-fed beef) by isotope analysis of their flesh and bone collagen. Similarly, marine fish contain more ^{13}C than freshwater fish, with values approximating the C_4 and C_3 plants respectively.

The ratio of carbon-13 and carbon-12 isotopes in these types of plants is as follows:

- C_4 plants: -16 to -10 ‰
- CAM plants: -20 to -10 ‰
- C_3 plants: -33 to -24 ‰

Limestones formed by precipitation in seas from the atmospheric carbon dioxide contain normal proportion of ^{13}C. Conversely, calcite found in salt domesoriginates from carbon dioxide formed by oxidation of petroleum, which due to its plant origin is ^{13}C-depleted.

The ^{14}C isotope is important in distinguishing biosynthetized materials from man-made ones. Biogenic chemicals are derived from biospheric carbon, which contains ^{14}C. Carbon in artificially made chemicals is usually derived from fossil fuels like coal or petroleum, where the ^{14}C originally present has decayed below detectable limits. The amount of ^{14}C currently present in a sample therefore indicates the proportion of carbon of biogenic origin.

Nitrogen isotopes

The ratio of $^{15}N/^{14}N$ tends to increase with trophic level, such that herbivores have higher nitrogen isotope values than plants, and carnivores have higher nitrogen isotope values than herbivores. Depending on the tissue being examined, there tends to be an increase of 3-4‰ at each trophic level. A number of other environmental and physiological factors can influence the nitrogen isotopic composition at the base of the food web (i.e. in plants) or at the level of individual animals. For example, in arid regions, the nitrogen cycle tends to be more 'open' and prone to the loss of nitrogen (^{14}N in particular), and concentrates soils and plants in ^{15}N ^{14}N. This leads to relatively high ä^{15}N values in plants and animals in hot and arid ecosystems relative to cooler and moister ecosystems.

The tissues and hair of vegans therefore contain significantly lower percentage of ^{15}N than the bodies of people who eat mostly meat. Isotopic analysis of hair is an important source of information for archaeologists, providing clues about the ancient diets; a terrestrial diet produces a different signature than a marine-based diet and this phenomenon has been used in analysing differing cultural attitudes to food sources.

Oxygen Isotopes

Oxygen comes in three variants, but the ^{17}O is so rare that it is very difficult to detect (~0.04% abundant). The ratio of $^{18}O/^{16}O$ in water depends on the amount of evaporation the water experienced (as ^{18}O is heavier and therefore less likely to vaporize). As the vapor tension depends on the concentration of dissolved salts, the $^{18}O/^{16}O$ ratio shows correlation on the salinity and temperature of water. As oxygen gets built into the shells of calcium carbonatesecreting organisms, such sediments prove a chronological record of temperature and salinity of the water in the area.

Oxygen isotope ratio in atmosphere varies predictably with time of year and geographic location; e.g. there is a 2% difference between ^{18}O-rich precipitation in Montana and ^{18}O-depleted precipitation in Florida Keys. This variability can be used for approximate determination of geographic location of origin of a material; e.g. it is possible to determine where a shipment of uranium oxide was produced. The rate of exchange of surface isotopes with the environment has to be taken in account.

APPLICATIONS OF ISOTOPE TRACERS IN CATCHMENT HYDROLOGY

The applications of environmental isotopes as hydrologic tracers in low temperature (< 40ºC) systems fall into two main categories:

- tracers of the water itself: *water isotope hydrology*
- tracers of the solutes in the water: *solute isotope biogeochemistry*.

Water Isotope Hydrology

Isotope Hydrology addresses the application of the measurements of isotopes that form water molecules: the oxygen isotopes (oxygen-16, oxygen-17, and oxygen-18) and the hydrogen isotopes (protium, deuterium, and tritium). These isotopes are ideal tracers of water sources and movement because they are integral constituents of water molecules, not something that is dissolved *in* the water like other tracers that are commonly used in hydrology (e.g., dissolved species such as chloride). Water isotopes can sometimes be useful tracers of water flowpaths, especially in groundwater systems where a source of water with a distinctive isotopic composition forms a "plume" in the subsurface.

In most low-temperature environments, stable hydrogen and oxygen isotopes behave conservatively in the sense that as they move through a catchment, any interactions with oxygen and hydrogen in the organic and geologic materials in the catchment will have a negligible effect on the ratios of isotopes in the water molecule. Although tritium also exhibits insignificant reaction with geologic materials, it does change in concentration over time because it is radioactive and decays with a half-life of about 12.4 years. The main processes that dictate the oxygen and hydrogen isotopic compositions of waters in a catchment are: (1) phase changes that affect the water above or near the ground surface (evaporation, condensation, melting), and (2) simple mixing at or below the ground surface.

Stable oxygen and hydrogen isotopes can be used to determine the contributions of old and new water to a stream (and to other components of the catchment) during periods of high runoff because the rain or snowmelt (new water) that triggers the runoff is often isotopically different from the water already in the catchment (old water). The sources of variability in the isotopic compositions of water in rain, snow, soil water, plants, and groundwater (respectively) and explain why the old and new water components often have different isotopic compositions. Tritium (^3H) is an excellent tracer for determining time scales for the mixing and flow of waters, and is ideally suited for studying processes that occur on a time scale of less than 100 years.

Solute Isotope Biogeochemistry

Isotope Biogeochemistry addresses the application of isotopes of constituents that are dissolved in the water or are carried in the gas phase. Isotopes commonly used in solute isotope biochemistry research include the isotopes of: sulfur, nitrogen, and carbon. Less commonly applied isotopes in geochemical research include those of: strontium, lead, uranium, radon, helium, radium, lithium, and boron.

Unlike the isotopes in the water molecules, the ratios of solute isotopes can be significantly altered by reaction with biological and/or geological materials as the water moves through the catchment. Although the literature contains numerous case studies involving the use of solutes (and sometimes solute isotopes) to trace *water* sources and flowpaths, such applications include an implicit assumption that these solutes are transported conservatively *with* the water. In a strict sense, *solute isotopes only trace solutes*. Solute isotopes also provide information on the reactions that are responsible for their presence in the water and the flowpaths implied by their presence.

Water isotopes often provide relatively unambiguous information about residence times and relative contributions from different water sources; these data can then be used to make hypotheses about water flowpaths. Solute isotopes can provide an alternative, independent isotopic method for determining the relative amounts of water flowing along various subsurface flowpaths. However, the least ambiguous use of solute isotopes in catchment research is tracing the relative contributions of potential solute sources to groundwater and surface water. Although there has been extensive use of carbon, nitrogen, and sulfur isotopes in studies of forest growth and agricultural productivity, solute isotopes are not yet commonly used for determining weathering reactions and sources of solutes in catchment research. This book attempts to remedy that situation.

Mixing

Isotopic compositions mix conservatively. In other words, the isotopic compositions of mixtures are intermediate between the compositions of the endmembers. Despite the awkward terminology (i.e., the notation and units of ‰) and negative signs, the compositions can be treated just like any other chemical constituent (e.g., chloride content) for making mixing calculations. For example, if two streams with known discharges (Q_1, Q_2) and known $d^{18}O$ values ($d^{18}O_1$, $d^{18}O_2$) merge and become well mixed, the $d^{18}O$ of the combined flow (Q_T) can be calculated from:

$$Q_T = Q_1 + Q_2$$

$$d^{18}O_T Q_T = d^{18}O_1 Q_1 + d^{18}O_2 Q_2.$$

Another example: any mixing proportions of two waters with known $d^{18}O$ and dD values will fall along a tie line between the compositions of the endmembers on a $d^{18}O$ vs. dD plot.

What is *not* so obvious is that on many types of X-Y plots, mixtures of two endmembers will not necessarily plot along lines but instead along *hyperbolic curves*. This is explained very elegantly by Faure (1986) using the example of $^{87}Sr/^{86}Sr$ ratios. The basic principle is that mixtures of two components that have different *isotope ratios* (e.g., $^{87}Sr/^{86}Sr$ or $^{15}N/^{14}N$) and different *concentrations* of the element in question (e.g., Sr or N) form hyperbolas when plotted on diagrams with coordinates of isotope ratios versus concentration. As the difference between the elemental concentrations of two components (endmembers) approaches 0, the hyperbolas flatten to lines. The hyperbolas are concave or convex depending on whether the component with the higher isotope ratio has a higher or lower concentration than the other component. Mixing hyperbolas can be transformed into a straight lines by plotting isotope ratios versus the inverse of concentration (1/C). Graphical methods are commonly used for determining whether the data support an interpretation of *mixing* of two potential sources or *fractionation of a single source.*

Implicit in such efforts is often the idea that mixing will produce a "line" connecting the compositions of the two proposed endmembers whereas fractionation will produce a "curve." However, both mixing and fractionation (in this case, denitrification) can produce curves (Mariotti et al., 1988), although both relations can look linear for small ranges of concentrations. However, the equations describing mixing and fractionation processes are different and under favorable conditions, the process responsible for the curve can be identified. This is because Rayleigh fractionations are *exponential* relations, and plotting d values versus the natural log of concentration will produce a straight line. If an exponential relation is not observed and a straight line is produced on a d vs 1/C plot, this supports the contention that the data are produced by simple mixing of two endmembers.

Isotopically Labeled Materials

Man-made materials with isotopic compositions that are not observed in nature are called "spiked" or isotopically labeled materials. There are many commercial suppliers of isotopically labeled liquids, gases, and solids — some with multiple-labeled atoms (e.g., water with unusual $^{18}O/^{16}O$ and $^{2}H/^{1}H$ ratios, or organic molecules with various percentages of the elements of specific functional groups labeled with uncommon isotopic compositions).

The most common watershed use of spiked tracers is for agricultural studies of plant uptake of nutrients. Other applications include whole-catchment experiments where labeled NH_4, NO_3, or SO_4 is sprinkled in artificial rain, and plot studies where labeled H_2O is applied to the land surface to make it easier to trace to movement of "new" water into the subsurface.

Materials can be enriched in either the common or less common isotope. Advantages of the former include low price, ready availability, and absence of potential contamination problems. The main disadvantage is that the lowest possible value for a material is -1000‰. In contrast, materials enriched in the less common isotope with values greater than $+10 \cdot 10^6$ ‰ are commonly available. Why the the lower limit of the permil scale is -1000‰ is illustrated by the following example for a water with no deuterium (i.e., all the hydrogen is protium):

$$d^2H = [(^2H / ^1H)_x / (^2H / ^1H)_s - 1)] \cdot 1000$$

$$d^2H = [(0/^1H)_x / (^2H / ^1H)_s - 1)] \cdot 1000 = (0 - 1) \cdot 1000 = -1000‰$$

The isotopic compositions in "labeled tracer" catalogs are generally in units of atom weight percent (at.%). For accurate conversion of these values to d values, one must know the R_s value of the appropriate standard used for that isotope. Unfortunately, the absolute R_s values are not known for all international standards; the average terrestrial abundance ratios can be used for rough estimates. For example, the d^{2}H value of a bottle of "95 at.% ^{2}H" water is calculated as follows:

$$d^2H = [(95/5) / (156 \cdot 10^{-6}) - 1)] \cdot 1000 = +122 \cdot 10^6 \text{ ‰}.$$

Although d values are additive for natural abundance studies, mass balance calculations for labeled materials should be done using fractional isotopic abundances where $F = R/(1 + R)$ and R is the ratio of isotopes of interest. For the general case where the *concentrations* of labeled material in the waters mixed together might be different (e.g., a water with 20 mg/L of 75 at.% ^{15}N-labeled NO_3 added to water with 5 mg/L of $NO_3$with a d^{15}N value of +2‰), the isotopic composition of the solute in the mixed solution is:

$$F_T C_T n_T = F_1 C_1 n_1 + F_2 C_2 n_2$$

where C is the concentration of the species of interest, n is the number of liters of solution, and the subscripts *T*, *1*, and *2* refer to the total, 1st, and 2nd waters, respectively.

Stable Isotopes in Geochemical Modeling

In chemical reaction modeling, usually several reaction models can be found that satisfy the data. For each model reaction path, calculations are used to predict the chemical and isotopic composition of the aqueous phase as well as the amounts of minerals dissolving or precipitated along a flow path. The power of the stable isotope technique in groundwater modeling lies in the fact that we have added one more thermodynamic component to our system for each isotope ratio that is measured (Plummer et al., 1983). These isotopic compositions can be used along with chemical data in geochemical mass balance and reaction path models (e.g., BALANCE, PHREEQE, NETPATH, etc.) to deduce geochemical processes, test hypotheses on hydrology and geochemical mechanisms, and eliminate possible reaction paths (Plummer et al., 1991).

For example, the d^{13}C of total dissolved inorganic carbon (DIC) is generally a function of the d^{13}C of the rocks and extent of reaction with the rocks in a system. Thus, d^{13}C can be a good indicator of which geochemical reactions are occurring. Sulfur is similar to carbon in this respect, and changes in d^{13}C along a flowpath sometimes can reflect reactions that also cause changes in d^{34}S (e.g., progressive calcite precipitation along a flowpath in response to degassing of CO_2 causes gypsum to dissolve). Changes in ^{14}C content along a flowpath are useful for indicating changes in residence time. On the other hand, there is little change in dD andd^{18}O of water during reactions with minerals along shallow, low-temperature flowpaths. Therefore, sulfur and carbon isotope data along a flowpath can sometimes be used to eliminate one or more plausible reaction models developed from chemical data, by comparing the observed changes in isotopic compositions with reaction progress along a flowpath. Other useful stable isotope tracers include d^{15}N and d^{18}O of nitrates and d^{34}S and d^{18}O of sulfates. Useful radiogenic isotopes include carbon-14, strontium-87, and various uranium-series isotopes.

Use of a Multi-isotope Approach for the Determination of Flowpaths

Flowpaths are the individual pathways contributing to surface flow in a catchment. These result from runoff mechanisms that include, but are not limited to, saturation-excess overland flow, Hortonian overland flow, near-stream groundwater ridging, hillslope subsurface flow through the soil matrix or macropores, and shallow organic-layer flow. Knowledge of hydrologic flowpaths in catchments is critical to the preservation of public water supplies and the understanding of the transport of point and non-point source pollutants (Peters, 1994). The need to incorporate flowpath dynamics is recognized as a key ingredient in producing reliable chemical models (Robson et al., 1992). In other words, if the model used gets the hydrology wrong, it is unlikely to correctly predict the geochemical response.

Stable isotopes such as ^{18}O and ^{2}H are shown throughout this book to be an improved alternative to traditional non-conservative chemical tracers because waters are often uniquely labeled by their isotopic compositions (Sklash and Farvolden, 1979), often allowing the separation of waters from different sources (e.g., "new" rain vs. "old" pre-storm water). However, studies have shown that flowpaths commonly cannot be identified to a high degree of certainty using d^{18}O or dD data and simple hydrograph separation techniques because waters within the same flowpath can be derived from several different sources (Ogunkoya and Jenkins, 1991).

One solution is to include alternative, independent isotopic methods for determining the relative amounts of water flowing along different subsurface flowpaths into hydrologic models. Reactive solute isotopes such as ^{13}C,^{34}S, and ^{87}Sr can provide valuable information about flowpaths (not water sources) useful for geochemical and hydrologic modeling precisely because they *can* reflect the reactions characteristic of and taking place along specific flowpaths. In many instances, the waters flowing along mineralogically distinctive horizons can be distinctively labeled by their chemical composition and by the isotopic compositions of solute isotopes like ^{13}C, ^{87}Sr, ^{34}S, ^{15}N, etc. For example, waters flowing through the soil zone often have d^{13}C values that are depleted in ^{13}C relative to deeper groundwaters because of biogenic production of carbonic acid in organic soils; these same shallow waters can also have distinctive Pb and Sr isotopic compositions.

SAMPLE COLLECTION, ANALYSIS, AND QUALITY ASSURANCE

Sampling Guidelines

Considerable field effort is often required to collect a sample that adequately represents the average composition of the medium being sampled, at the time it is sampled. For small streams, this can be as simple as collecting water as it flows over a weir or rock ledge. For large rivers, lakes, soils, and organisms, mass-integrated composites may be required. Adequate coverage of this vital topic is beyond the scope of this chapter. The reader is advised to look at the references given in subsequent chapters, or consult colleagues who routinely collect such samples.

Below is a potpourri of guidelines and suggestions related to collecting, bottling, and preserving samples for analysis of the most commonly-used environmental isotopes. The reader should keep in mind that the optimum methods often depend on the laboratory chosen for analysis and their preferred preparation methods, and should always inquire before planning the field campaign. Collection of duplicates is alway advisable — in case of breakage of samples during transport and to use as checks of the reproducibility of the laboratory (i.e., submit 5-10% of these as "blind duplicates," with different sample ID numbers than their duplicates).

$d^{18}O$/ ^{2}H of water

Natural waters are easy to collect. The water sample is put in a clean dry bottle, which is filled almost completely to the top, and capped tightly. The main objective is to protect the sample from evaporation and exchange with atmospheric water vapor. Samples should not be filtered unless they contain oil (e.g., mineral oil added to rain collectors to help prevent evaporation) or contain abundant particulate matter. Bottle rinsing, chilling, and addition of preservatives are unnecessary. Freezing does not affect the composition of the water but can break the bottles in transit; for this reason, many users prefer plastic bottles. Our experience suggests that caps with conical plastic inserts (e.g., "poly-seal" caps) are the most reliable, followed by teflon-lined caps. For extended storage, use of glass bottles and waxing of the caps is advisable. Sample-size is lab-dependent; typical volumes range from 10-60 mL. In some laboratories, samples as small as a few μL can be analyzed.

Determinations of both hydrogen and oxygen isotope ratios are usually made on the same bottle of water. It is wise to collect many more samples than one can afford to analyze at the present; samples have a long shelf life if bottled correctly, and can be archived for future analysis. One should make sure that the laboratory chosen to analyze the samples normalizes their values according to IAEA guidelines (Coplen, 1994), and reports values relative to

VSMOW. If the samples are saline, one should check whether the lab is preparing samples by an equilibration or quantitative-conversion method. Waters with high contents of volatile organic matter may require distillation.

For many purposes, especially hydrograph separations, analysis for all samples for both oxygen and hydrogen isotopes is unnecessary because of the high correlation coefficient between these isotopes. A sensible alternative is to have some smaller percentage analyzed for both isotopes, either initially or after the data for the first isotope are evaluated. For hydrograph studies in arid environments or studies that involve evaporated water in ponds or wetlands, analysis of samples for both isotopes is probably advisable. Because most labs have fewer problems analyzing waters for $d^{18}O$ than for $d^{2}H$, if the samples are not analyzed in duplicate and will only be analyzed for one isotope, it is usually better to choose $d^{18}O$.

Solid and vapor samples are more difficult to collect for $d^{18}O$ and $d^{2}H$. Snow and ice samples can be collected in tightly sealed bags or jars, melted overnight, and then poured into bottles. Plant and soil samples should be collected in air-tight containers matched to the sample size. Common procedures include waxing of soil cores, use of heat-sealed bags, or insertion into tiny tree-core-size vials. Water vapor samples are collected by pumping vapor through a cold-trap where the vapor is quantitatively retained.

Tritium

The amount of water needed for tritium analysis depends on the age of the water (old waters contain little tritium) and the sensitivity of analysis needed. Typical sample sizes range from 10 mL to 1L. Samples are collected in unrinsed glass or high-density polyethylene bottles and should not be filtered. The bottles should then be sealed and returned to the laboratory for analysis. The collection date should be noted on the bottle to obtain an accurate determination of the tritium concentration for the time of collection.

$d^{13}C$ and ^{14}C of Dissolved Inorganic Carbon

There are two main methods in common use for the collection of DIC (dissolved inorganic carbon) for the measurement of ^{13}C or ^{14}C, depending on which of two laboratory preparation methods is being used: gas stripping or carbonate precipitation.

Both preparation methods insure quantitative removal of the DIC and provide a $d^{13}C$ or ^{14}C value for total DIC. Analysis for $d^{13}C$ generally requires 10-100 μM of carbon. Analysis of^{14}C by conventional beta-counting methods requires as much as 1 g of C; analysis by AMS usually requires about 1 mg of C.

For laboratories that use a gas-stripping method to extract the CO_2, samples are usually collected in sample-rinsed glass bottles with septa-caps, or in vessels with stopcocks or valves. Such samples should be filtered to remove particulate carbon, and perhaps poisoned (using mercuric chloride, acid, or organic biocide) to prevent biological activity; the bottles should be kept chilled until analyzed to prevent biological fractionations.

The alternative technique is the precipitation method. Samples should be pre-filtered if there might be suspended carbonate particulate material in the water. The carbonate is precipitated by adding a strongly basic solution of strontium or barium chloride (Gleason et al., 1969) to the sample in a sample-rinsed bottle. The base increases the pH to 10-11 where all the inorganic carbon is CO3-2, and the Ba or Sr precipitates all the DIC in the water. This reagent and the treated samples must be protected against contamination by atmospheric CO2. Glass bottles are best because CO2 diffuses through most plastic bottles. Bottles should have poly-seal caps that are taped securely. Bottles should be individually wrapped in bubble paper and shipped in insulated boxes or coolers filled with artificial "peanuts" to insure against breakage.

d^{15}*N of Dissolved Inorganic Nitrogen*

A number of different preparation methods are in common use; inquire what collection method is preferred by the contract laboratory for their particular preparation method. In particular, it is important to verify that the laboratory is accustomed to analyzing natural abundance samples. Laboratories that primarily analyze agricultural samples often use methods that are appropriate for labeled (^{15}N-spiked) samples but have unacceptable analytical precisions for natural abundance studies.

Check that the laboratory has a good track record for natural samples. Samples can be analyzed for the d^{15}N of ammonium and/or nitrate; analysis of total nitrogen is probably worthless. Generally, samples are filtered through 0.1 micron filters, put in rinsed glass bottles, poisoned (with sulfuric acid, mercuric chloride, or chloroform), chilled or frozen, wrapped in insulating packing material, and sent to the laboratory in ice chests. Sample-size requirements are in the range of 10-100 μM of N. Nitrate samples can also be analyzed for d^{18}O in a few laboratories.

An alternate method is to concentrate the NO_3 or NH_4 on anion or cation exchange resins (Garten, 1992; Silva et al., submitted; Chang et al., in review). Collection of nitrate on anion exchange resins eliminates the need to send large quantities of chilled water back to the laboratory, eliminates the need for hazardous preservatives, makes it easier to archive samples, and allows analysis of extremely low-nitrate waters.

d^{34}S of Dissolved Sulfate

Depending on the sulfate concentration, samples are filtered directly into glass bottles or are first pre-concentrated on an exchange resin. Sulfate from dilute waters should be collected on ion exchange resin in the field if the concentration of sulfate in the water is believed to be less than 20 mg/L. Similar to collection methods for NO_3 or NH_4 on ion exchange resins, collection of sulfate on exchange resins avoids problems of incomplete precipitation of $BaSO_4$ in dilute samples, eliminates the need to send large quantities of chilled water back to the lab, eliminates the need for hazardous preservatives, makes it easier to archive samples, and allows analysis of extremely low-sulfate waters.

Low-sulfate water samples are first acidified before passing through ion exchange columns. The sulfate is then eluted from the resin using a relatively small volume of concentrated barium chloride solution. The final volume of the solution is much less than that of the original water sample and the sulfate from the sample is thus concentrated in this much smaller volume (generally 10-500 μM of SO_4 is required). The solution is reacidified and sulfate is precipitated by adding $BaCl_2$. $BaSO_4$ is then collected by filtration and analyzed for $d^{34}S$. Sulfate can also be analyzed for $d^{18}O$ in some laboratories. Large quantities of sulfate can also be analyzed for ^{35}S, a natural radioisotope with a half-life of 87 days, using liquid scintillation counting.

C, H, N, O, and S Isotopes of Solid Samples

Solid organic and inorganic samples (e.g., animals, plants, minerals, and soils) and liquids (such as oils) can also be analyzed for their isotopic composition. Particulate matter in water can be captured on fiberglass filters and processed similar to methods used for other solid samples. Requirements for solid samples are similar to the requirements for solute samples of the same element (i.e., 1-100 μM of the element of interest). Biologically labile samples (e.g., leaves, fish, manure) should be kept cold until processed. Freeze-drying is an ideal means for preserving the samples; air-drying results in loss of volatile organic matter and probably some isotopic fractionation.

Lithogenic (Metals and Semi-metals) Isotopes

The sample size is dependent on the species being analyzed. Analysis of Sr, Li, or B requires a minimum of 1 μg; Pb and Nd require a minimum of 0.1 μg. Aqueous samples should be filtered; 0.1 micron filters are best for Nd and Pb, and 0.45 micron filters are best for Sr, Li, and B (Thomas D. Bullen, pers. comm. 1997). Aqueous samples are collected in rinsed plastic bottles and acidified to pH ~ 2 using Ultrex HNO_3. Blanks should be sent to the laboratory along with your samples, including the triple distilled water used for filtering and the clean water run through the processing equipment. One must be careful about possible contamination with lithium grease, borate soaps or detergents, and strontium chloride reagents.

Analytical Methods and Instrumentation

Stable isotopes are analyzed either on *gas-* or *solid-source* mass spectrometers, depending on both the masses of the isotopes and the existence of appropriate gaseous compounds stable at room temperature. Radioisotopes can be analyzed by counting the number of disintegrations per unit time on gamma ray or beta particle counters, or analyzed on mass spectrometers.

Gas-source Mass Spectrometers

Many methods are used to prepare gases for C, H, N, O, and S (CHNOS) stable isotope content, but in all the cases the basic steps are the same. Sample preparation involves the quantitative conversion or production of pure gas from solely the compound of interest, cryogenic or chromatographic purification of the gas, introduction of the gas into the mass spectrometer, ionization to produce positively charged species, dispersion of different masses in a magnetic field, impaction of different masses on different collector cups, and measurement of the ratios of the isotopes in the ionized gas. In general, hydrogen is analyzed as H_2, oxygen and carbon are both analyzed as CO_2, nitrogen is analyzed as N_2, and sulfur is usually analyzed as SO_2. The analytical precisions are small relative to the ranges in values that occur in natural earth systems. Typical one standard deviation analytical precisions for oxygen, carbon, nitrogen, and sulfur isotopes are in the range of 0.05 to 0.2‰; typical precisions for hydrogen isotopes are poorer, from 0.2 to 1.0‰, because of the lower ^2H:^1H ratio.

Although the topic is rarely discussed, the activity coefficients of isotopic species are not all equal to 1 (i.e., the isotope *concentration* of a sample is not necessarily equal to the isotope *activity*). The activity coefficient for a particular isotope can be positive or negative, depending on solute type, molality, and temperature. The isotopic compositions of waters and solutes can be significantly affected by the concentration and types of salts because the isotopic compositions of waters in the hydration spheres of salts and in regions farther from the salts are different. In general, the only times when it is important to consider isotope activities is for low pH, high SO_4, and/or high Mg brines because the activity and concentration d values of these waters (d_a and d_c) are significantly different. For example, the difference (dD_a - dD_c) between the activity and concentration d values for sulfuric acid solutions in mine tailings is about +16‰ for 2 molal solutions. For normal saline waters (e.g., seawater), the activity coefficients for $d^{18}O$ andd^2H are essentially equal to 1.

Virtually all laboratories report $d^{18}O$ activities (not concentrations) for water samples. The d^2H of waters may be reported in either concentration or activity d values, depending on the method used for preparing the samples for analysis. Methods that involve *quantitative conversion* of the H in H_2O to H_2, produce d_c values. Methods that *equilibrate* H_2O with H_2 (or H_2O with CO_2) produce d_a values. "Equilibrate" in this case means letting the liquid and gas reach isotopic equilibrium at a constant, known temperature. To avoid confusion, laboratories and research papers should always report the method used.

Most conventional CHNOS mass spectrometers are *dual inlet* machines that have both a sample and a standard inlet or introduction port. In such instruments, the ratios of the isotopes of interest (e.g., $^{13}C/^{12}C$) in the sample gas are measured relative to the same ratios in a gaseous standard that is analyzed more-or-less simultaneously. Such instruments usually have either "double collectors" or "triple collectors," meaning that either two or three masses of the ionized gas can be measured simultaneously. For example, N_2 contains three species: $^{14}N^{14}N$, $^{14}N^{15}N$ and $^{15}N^{15}N$ (i.e., masses 28, 29, and 30). A triple-collecting mass spectrometer would measure the abundances of all these species relative to the abundances in a gaseous standard introduced through the "standard" inlet. A double-collecting mass spectrometer would only measure the 28 and 29 masses (actually m/e is measured since the molecules are ionized, with positive charges).

Another type of stable isotope mass spectrometer is the so-called *continuous flow* mass spectrometer. Such instruments may lack a dual inlet, and usually have triple collectors. These instruments represent a "marriage" of chromatography and mass spectrometry, and are similar to conventional organic mass spectrometers in that gas samples are introduced into the mass spectrometer within a stream of helium gas, usually from an automated sample preparation unit (e.g., an elemental analyzer or gas chromatograph). In general, the analytical precision available for continuous flow mass spectrometers is slightly poorer than with conventional methods, but this may change in the next few years. The main advantage of the continuous flow method is that such instruments are very easily combined with various on-line preparation systems, dramatically lowering the manpower cost of isotope analyses.

Solid-source Mass Spectrometers

Elements analyzed as solids (e.g., strontium, lithium, boron, lead, etc.) are prepared by precipitating selected compounds on wire filaments, loading the filaments into the source of a thermal ionization (solid-source) mass spectrometer, ionizing the compounds to produce gases (negative or positive charged), and measuring the abundances of selected isotopes in the gas on multiple collectors. Some light-mass solids (e.g., boron and lithium) are reported in the standard d units. Generally, the heavier-mass elements are reported in terms of the relative abundances of two isotopes (e.g., $^{207}Pb/^{206}Pb$); however, strontium isotope abundances ($^{87}Sr/^{86}Sr$) are occasionally reported in d notation relative to some arbitrary standard. Solid-source mass spectrometry has been shown to give a more accurate analysis of certain radium and uranium isotopes that conventionally were measured by decay counting methods.

Gas and Liquid Scintillation Counters

Radioactive isotopes can be measured by a number of methods, depending on the mass, abundance, type of decay involved, accuracy desired, and money available. Some, of course, can be analyzed on solid source mass spectrometers (e.g., uranium-series isotopes). Otherwise, radioisotopes are analyzed on liquid scintillation counters (LSC) and gas proportional counters (both with enrichment), and on accelerator mass spectrometers. Liquid scintillation and gas proportional systems are the most common systems used for light isotopes with beta decays. Gas proportional counting usually requires that the isotope being analyzed form a suitable counting gas, so that elements with high electronegativities, such as chlorine and sulphur, are not suitable for this type of analysis. The two isotopes most commonly used in hydrology, tritium and ^{14}C, have generally been analyzed using liquid scintillation or gas proportional counting. Radon is analyzed either by gas Geiger or proportional counting in the field, or sent to a laboratory for liquid scintillation counting, depending on the accuracy desired. For isotopes that decay by gamma and alpha emission, and beta emissions where the target isotope cannot be reduced to a suitable chemical form for LCS or gas counting, the use of solid scintillation-counting using crystals, or more advanced systems like lithium-germanium drift counters, have been utilized.

Accelerator Mass Spectrometry

Accelerator mass spectrometers (AMS), sometimes called "tandem" accelerators, are very large (> 10m), expensive, high-resolution, mass spectrometers (with either gas or solid-sources) that accelerate charged particles through very high (mega-volt) electrical fields to separate different isobars and isotopes. These instruments can analyze some radioactive species more rapidly, with greater accuracy, and/or with much smaller sample sizes (e.g., mg rather than g samples) than previous counting methods. For example, tritium can now be analyzed using the helium ingrowth method, although it frequently requires long delays (6 months) to accrue enough ^{3}He to obtain an accurate analysis. AMS has become the method of choice for some isotopes, such as ^{14}C, ^{36}Cl and ^{129}I. It will give accuracies close to those obtained by traditional methods, and samples can be analyzed much more rapidly by AMS.

Quality Assurance of Contract Laboratories

How does one find a good contract laboratory for analyzing samples? Choices include university laboratories, private commercial companies, and government laboratories that can accept contract (or collaborative) work. A primary selection criterion should be that the laboratory has been making the desired type of analysis for several years on a routine basis (e.g., samples submitted to some university laboratories may be analyzed by temporary student help, who do not perform analyses on a routine basis). Make inquiries among colleagues about the long-term track record of the laboratory. Good laboratories have active QA/QC programs, with documentation generally available on request. In our opinion, the laboratory should analyze about 5-15% of the samples in duplicate, as an internal verification that "everything" is operating correctly. Furthermore, laboratories with automated preparation systems and computer-controlled data management systems probably produce better and more reliable data on a long-term basis than laboratories where everything is done manually. The reader is cautioned to beware of bargains (*caveat emptor*); quality work usually costs more than the average price. Furthermore, the potential long-term cost of wrong interpretations, due to bad data, should be factored into the total cost of the analyses when evaluating laboratory choices.

One should also consider collecting duplicates in case the sample bottle is broken or lost in transit. Most laboratories routinely analyze each sample only once; if high precision data are required, either request duplicate analysis of each sample (and triplicates if the duplicates do not agree within some predetermined range) or send in "blind" duplicates. Sending in 10-15% blind duplicates is advisable, in any case. If any result seems questionable, immediately request a repeat. Most laboratories keep analyzed samples for a couple months before discarding them and will reanalyze modest numbers of samples at no additional cost. For water samples, immediately plot the data on a dD vs. $d^{18}O$ diagram; outliers, especially ones that plot appreciably above the GMWL, should be viewed with skepticism and possibly reanalyzed. Few natural processes produce waters that plot significantly above the GMWL; exceptions include methanogenesis in landfills and silicate hydrolysis.

STABLE ISOTOPE FRACTIONATION

Fractionation Accompanying Chemical Reactions and Phase Changes

The strength of chemical bonds involving different isotopic species will usually be different. Molecules containing heavy isotopes are more stable (i.e., have a higher dissociation energy) than molecules with lighter isotopes. Hence, isotopic fractionations between molecules can be explained by differences in their *zero point energies* (ZPE). For example, there is about a 2 kcal/mole difference in ZPE associated with the breaking of the H-H bond compared to the D-D bond. Hence, H-H bonds are broken more easily and D-D bonds are more stable. Chemical reaction rates where such a bond is broken will also show an isotope effect. These are quantum effects that become appreciable at low temperatures and disappear at higher temperatures. The energy differences associated with isotope effects are about 1000 times smaller than the DG for chemical reactions, and hence cannot be the driving force for chemical equilibrium.

Equilibrium Fractionations

Equilibrium isotope-exchange reactions involve the redistribution of isotopes of an element among various species or compounds (in a strict sense, this only occurs in a closed, well-mixed system at chemical equilibrium). At isotopic equilibrium, the forward and backward reaction rates of any particular isotope are identical. This does not mean that the isotopic compositions of two compounds at equilibrium are identical, but only that the ratios of the different isotopes in each compound are constant for a particular temperature.

During equilibrium reactions, the heavier isotope generally preferentially accumulates in the species or compound with the higher oxidation state. For example, sulfate becomes *enriched* in ^{34}S relative to sulfide (i.e., has a more positive various materials $d^{34}S$ value); consequently, the residual sulfide becomes *depleted* in various materials $d^{34}S$. As a "rule of thumb," among different phases of the same compound or different species of the same element, the more dense the material, the more it tends to be enriched in the heavier isotope. For example, for the various phases of water, at equilibrium, $d^{18}O_S > d^{18}O_L > d^{18}O_V$. Also, the $d^{13}C$ and $d^{18}O$ values of $CO_2 < HCO_3^- < CaCO_3$.

During phase changes, the ratio of heavy to light isotopes in the molecules in the two phases changes. For example, as water vapor condenses in rain clouds (a process typically viewed as an equilibrium process), the heavier water isotopes (^{18}O and ^{2}H) become enriched in the liquid phase while the lighter isotopes (^{16}O and ^{1}H) remain in the vapor phase. In general, the higher the temperature, the less the difference between the equilibrium isotopic compositions of any two species (because the differences in ZPE between the species become smaller). The fractionation associated with the equilibrium exchange reaction between two substances *A*and *B* (i.e., the fractionation of *A* relative to *B*) can be expressed by use of the isotope fractionation factor a (alpha):

$a_{A\text{-}B} = R_A / R_B$

where R = the ratio of the heavy isotope to the lighter isotope (i.e., D/H, $^{18}O/^{16}O$, $^{34}S/^{32}S$, etc.) in compounds *A* and *B*.

The value of such an *equilibrium* fractionation factor can be calculated on the basis of spectral data of the isotopic molecular species, at least for simple molecules. The a values generally differ by just a few percent from the equal-energy value of 1.00, except for exchange reactions involving hydrogen isotopes where a values may be as large as 4 at room temperature. The sign and magnitude of a are dependent on many factors, of which temperature is generally the most important. Other factors include chemical composition, crystal structure, and pressure.

The equilibrium fractionation factors ($a_{l\text{-}v}$) for the water liquid-vapor phase transition are 1.0098 and 1.084 at 20ºC and 1.0117 and 1.111 at 0ºC for ^{18}O and ^{2}H, respectively. In both cases, $a_{l\text{-}v} > 1$, which means that the first phase (the liquid water) is "heavier" than the second phase (e.g., for $a_{l\text{-}v} = 1.0098$, the $d^{18}O$ of water is +9.8‰ higher than the $d^{18}O$ value of vapor at equilibrium). For the ice-water transition (0ºC), the values are 1.0035 and 1.0208, respectively.

A useful equation that relates d values and fractionation factors is:

$a_{A\text{-}B} = (1000 + d_A) / (1000 + d_B)$.

Other common formulations for fractionation factors include:

$a^* = 1/a = a_{B\text{-}A} = R_B / R_A$

and

$e_{A\text{-}B} = (a_{A\text{-}B} - 1) \cdot 1000$

For small values of e (epsilon), $e_{A\text{-}B} \sim d_A - d_B$. For example, if $d_B = +10‰$ and if $a_{A\text{-}B} = 1.020$, then = 20‰ and $d_A \sim +30‰$. The difference in isotopic composition between two species *A* and *B* is defined as:

$e_{A\text{-}B} \sim d_A - d_B \sim 1000 \ln a_{A\text{-}B}$.

Fractionation factors are commonly expressed as "$10^3 \ln a$" because this expression is a very close approximation to the permil fractionation between the materials (e), especially for the values a of near to unity typical of most elements of interest (O'Neil, 1986), and because the value "$10^3 \ln a$" is nearly proportional to the inverse of temperature (1/T) at low temperatures (ºK). Graphical plots of the temperature dependency of a are typically given as $10^3 \ln a$ versus 1/T (Friedman and O'Neil, 1977).

Kinetic Fractionations

Chemical, physical, and biological processes can be viewed as either reversible *equilibrium* reactions or irreversible unidirectional *kinetic* reactions. In systems out of chemical and isotopic equilibrium, forward and backward reaction rates are not identical, and isotope reactions may, in fact, be unidirectional if reaction products become physically isolated from the reactants. Such reaction rates are dependent on the ratios of the masses of the isotopes and their vibrational energies, and hence are called kinetic isotope fractionations.

The magnitude of a kinetic isotope fractionation depends on the reaction pathway, the reaction rate, and the relative bond energies of the bonds being severed or formed by the reaction. Kinetic fractionations, especially unidirectional ones, are usually larger than the equilibrium fractionation factor for the same reaction in most low-temperature environments. As a rule, bonds between the lighter isotopes are broken more easily than equivalent bonds of heavier isotopes. Hence, the light isotopes react faster and become concentrated in the products, causing the residual reactants to become enriched in the heavy isotopes. In contrast, reversible equilibrium reactions can produce products heavier or lighter than the original reactants.

Many reactions can take place either under purely equilibrium conditions or be affected by an additional kinetic isotope fractionation.

For example, although isotopic exchange between water and vapor can take place under more-or-less equilibrium conditions (i.e., at 100% humidity when the air is still and the system is almost chemically closed), more typically the system is out of chemical equilibrium (i.e., $< 100\%$ humidity) or the products become partially isolated from the reactants (e.g., the resultant vapor is blown downwind). Under these conditions, the isotopic compositions of the water and vapor are affected by an additional kinetic isotope fractionation of variable magnitude.

Isotope fractionation factors can be defined as:

$$a = R_p / R_s$$

where R_p and R_s are the ratios of the heavy to light isotope in the *product* and *substrate* (reactant), respectively. An isotope enrichment factor, e, can be defined as:

$$e_{p\text{-}s} = (a - 1) \cdot 1000.$$

If the reactant concentration is large and fractionations are small,

$$e_{p\text{-}s} \sim D = d_p - d_s \ ($$

where D (del) is another term for the enrichment factor. The kinetic fractionations are the same as for equilibrium fractionations, except for the differences in subscripts. One should be especially careful with the superscripts, subscripts, and units of all fractionation factors; different authors may define them differently. The use of $_p$ and $_s$ (or $_r$) for kinetic fractionations like the unidirectional nature of these reactions.

The same formulations apply not only when part of the system is removed by a chemical or biological reaction, but also when material escapes by diffusion or outflow (e.g., by effusion through an aperture). In the latter cases the term *transport* fractionation factor may be preferred. The transport fractionation, like the *equilibrium* factors, is temperature dependent. However, unlike true *kinetic* fractionation factors, which can be quite appreciable, transport fractionations have only slight (positive) temperature coefficients

The Rayleigh Equation

The isotopic literature abounds with different approximations of the Rayleigh equations, including the three equations below. These equations are so-named because the original equation was derived by Lord Rayleigh (pronounced "raylee") for the case of *fractional distillation of mixed liquids*. This is an exponential relation that describes the partitioning of isotopes between two reservoirs as one reservoir decreases in size. The equations can be used to describe an isotope fractionation process if: (1) material is continuously removed from a mixed system containing molecules of two or more isotopic species (e.g., water with ^{18}O and ^{16}O, or sulfate with ^{34}S and ^{32}S), (2) the fractionation accompanying the removal process at any instance is described by the fractionation factor a, and (3) a does not change during the process. Under these conditions, the evolution of the isotopic composition in the residual (reactant) material is described by:

$$(R / R_{o}) = (X_1 / X_{1o})^{a-1}$$

where R = ratio of the isotopes (e.g., $^{18}O/^{16}O$) in the reactant, R_o = initial ratio, X_1 = the concentration or amount of the more abundant (lighter) isotope (e.g.,^{16}O), and X_{1o} = initial concentration. Because the concentration of $X_1 >> X_h$, X_1 is approximately equal to the amount of original material in the phase. Hence, if $f = X_1/X_{1o}$ = fraction of material remaining, then:

$$R = R_o f^{(a-1)}.$$

Another form of the equation in d-units is:

$$d \sim d_o f^{(a-1)}$$

which is valid for a values near 1, d_o values near 0, and values less than about 10.

In a strict sense, the term "Rayleigh fractionation" should only be used for chemically *open* systems where the isotopic species removed at every instant were in thermodynamic and isotopic equilibrium with those remaining in the system at the moment of removal. Furthermore, such an "ideal" Rayleigh distillation is one where the reactant reservoir is finite and well mixed, and does not re-react with the product (Clark and Fritz, 1997). However, the term "Rayleigh fractionation" is commonly applied to equilibrium closed systems and kinetic fractionations as well because the situations may be computationally identical.

Isotopic Fractionation in Open and Closed Systems

The Rayleigh equation applies to an open system from which material is removed continuously under condition of a constant fractionation factor. However, such processes can proceed under different boundary conditions, even when the fractionation factors are the same. One such system is the so-called "closed" system (or 2-phase equilibrium model), where the material removed from one reservoir accumulates in a second reservoir in such a manner that isotopic equilibrium is maintained throughout the process (Gat and Gonfiantini, 1981). An example is the condensation of vapor to droplets in a cloud where there is continuous exchange between the isotopes in the vapor and water droplets.

The isotope enrichment achieved can be very different in closed vs. open systems. The changes in the $d^{18}O$ of water and vapor during evaporation (an *open-system* process) where the vapor is continuously removed (i.e., isolated from the water) with a constant fractionation factor $a_{l\text{-}v}$ = 1.010 (i.e., the newly formed vapor is always 10‰ lighter than the residual water). As evaporation progresses (i.e., f—>O), the $d^{18}O$ of the remaining water (solid line A), becomes heavier and heavier. The $d^{18}O$ of the instantaneously formed vapor (solid line B) describes a curve parallel to that of the remaining water, but lower than it (for all values of f) by the precise amount dictated by the fractionation factor for ambient temperature, in this case by 10‰. For higher temperatures, the a value would be smaller and the curves closer together. The integrated curve, giving the isotopic composition of the accumulated vapor thus removed, is shown as solid line C. Mass balance considerations require that the isotope content of the total accumulated vapor approaches the initial water $d^{18}O$ value as f—>O; hence, any process should be carried out to completion (with 100% yield) to avoid isotopic fractionation.

The $d^{18}O$ of vapor (E) and water (D) during equilibrium evaporation in a *closed* system (i.e., where the vapor and water are in contact for the entire phase change). Note that the $d^{18}O$ of vapor in the *open* system where the vapor is continuously removed (line B) is always heavier than the $d^{18}O$ of vapor in a *closed* system where the vapor (line E) and water (line D) remain in contact. In both cases, the evaporation takes place under equilibrium conditions with a = 1.010, but the cumulative vapor in the closed system *remains* in equilibrium with the water during the entire phase change. As a rule, fractionations in a true "open-system" Rayleigh process create a much larger range in the isotopic compositions of the products and reactants than in closed systems. This is because of the lack of back reactions in open systems. Natural processes will produce fractionations between these two "ideal" cases.

Other non-equilibrium fractionations may behave like Rayleigh fractionations in that there may be negligible back reaction between the reactant and product, regardless of whether the system is open or closed, because of kinetics. Such fractionations typically result in larger ranges of composition than for equivalent equilibrium reactions. An example of this process is biologically mediated denitrification (reduction) of nitrate to N_2 in groundwater; the N_2 is lost so it can't re-equilibrate with the nitrate, even if there was a back reaction by this organism, which there isn't. Rayleigh-type fractionations affect the compositions of residual substrate, instantaneous product, and cumulative product (curved lines) during a *closed*-system kinetic reaction (e.g., denitrification, uptake of N by plants, or nitrification). Note that at all times, the d values of instantaneous product are " ‰" less than the corresponding d values of residual substrate. The parallel straight lines are the compositions for an *open*-system kinetic reaction where the supply of substrate is infinite and, hence, is not affected by the conversion of some substrate to product with a constant fractionation of e.

The *boundary* lines are drawn between the system being studied and the rest of the universe. In the case of the *equilibrium* fractionations, "open" means that the product, once formed at equilibrium, escapes to outside the system and does not interact again with the residual substrate (and, consequently, is no longer in equilibrium with the substrate). And "closed" means that the reactant and product remain in close contact, in their own closed (finite) system during the entire reaction, so that the two reservoirs are always in chemical and isotopic equilibrium. The supply of substrate is infinite (which it can't be in a closed system). The use of "closed" for kinetic reactions suggests that there is a limited supply of reactant, which is undergoing irreversible, quantitative, conversion to product in an isolated system.

Thusfar, a constant fractionation factor was assumed to apply throughout the process. However, this is not always the case. For example, rainout from an air mass is usually the result of a continuous cooling of the air parcel. The cooling increases the fractionation factor for the vapor-to-water (or vapor-to-ice) transition. Another conspicuous example of a changing "effective" fractionation factor is that of the evaporation of water from a surface water body to the atmosphere. As will be shown, the change in this situation is the result of the changing conditions (in this case, of the isotopic gradient) at the water-atmosphere boundary, rather than a change of the fractionation factors themselves.

Almost everyone finds the Rayleigh equations a bit confusing. Hence, we will now give some examples of how to calculate open- and closed-system fractionations, and how they affect the compositions of the residual substrate and the newly formed products of a reaction. Because much of the book focuses on *water* and its isotopes, we will demonstrate how to apply the Rayleigh equations by using the fractionations during water phase changes (i.e., during the condensation of vapor and the evaporation of water) as examples.

Condensation of Water

The isotopic composition of moisture in the marine atmosphere is controlled by the air-sea interaction processes as described by Craig and Gordon (1965), Merlivat and Jouzel (1979), and others. As air masses move across continents and lose water by rainout, they become *depleted* in the heavy isotopic species ($H_2{}^{18}O$ and HDO) because the liquid phase is *enriched* in the heavy isotopic species relative to the vapor phase. The evolution of the isotopic composition is adequately described by a Rayleigh process (in this case it is condensation) in those cases where rainout is the only factor in the atmospheric-moisture budget (Dansgaard, 1964; Gat, 1980). A Rayleigh fractionation plot for condensation would be very similar, except that all the curves would bend down instead of up because the residual vapor and water condensed would become progressively *lighter* over time, not heavier (as they do for evaporation).

When the isotopic compositions of precipitation samples from all over the world are plotted relative to each other on $d^{18}O$ versus dD plots, the data form a linear band of data that can be described by the equation (Craig, 1961):

$$dD = 8\ d^{18}O + 10$$

and is called the *Global Meteoric Water Line* (GMWL) or just the *MWL*, or even the *Craig Line*. The slope is 8 (actually, different data sets give slightly different values) because this is approximately the value produced by equilibrium Rayleigh condensation of rain at about 100% humidity. The value of 8 is also close to the ratio of the equilibrium fractionation factors for H and O isotopes at 25-30ºC. At equilibrium, the d values of the rain and the vapor both plot along the MWL, but separated by the ^{18}O and ^{2}H enrichment values corresponding to the temperature of the cloud base where rainout occurred.

The y-intercept value of 10 in the GMWL equation is called the *deuterium excess* (or d-excess, or d parameter) value for this equation. The term only applies to the calculated y-intercept for sets of meteoric data "fitted" to a slope of 8; typical d-excess values range from 0 to 20. The fact that the intercept of the GMWL is 10 instead of 0 means that the GMWL does not intersect $d^{18}O = dD = 0$, which is the composition of average ocean water (VSMOW). The GMWL does not intersect the composition of the ocean, the source of most of the water vapor that produces rain, because of the 10‰ kinetic enrichment in D of vapor evaporating from the ocean at an average humidity of 85%.

The Rayleigh law is formulated in approximate differential form and using notation as:

$$d\mathit{d} \sim e^{*} \cdot d\ln f$$

where $f = N_f/N_0$ is the fraction of remaining water (N_0 and N_f being the water content of the air mass before and after the rain, respectively) so that $(N_0 - N_f)$ is the total water loss (rainout) from the air mass. The term $e^*$is related to a^*, the unit equilibrium isotope fractionation factor between water and its vapor at the ambient near-surface air temperature, as follows:

$$e^+ = (a^+ - 1) \cdot 10^3$$

Note that this equation is the same, except for the superscripts. Why the change? Because of some historical choices made to simplify mathematical expressions.

The equilibrium fractionation factor a between liquid and vapor can be defined in two ways, which are mathematical inverses: $a = R_l / R_v$ or $a = R_l / R_v$, where R_l and R_v are the isotopic ratios of the liquid and vapor, respectively. However, Craig and Gordon (1965) defined equilibrium fractionation factors such that $a^+ = 1/a^*$, so that $a^+ = R_l / R_v > 1$ and $a^* = R_v / R_l < 1$ (and, consequently, $e^+ > 0$ and $e^+ - e^*$). This usage has become traditional when discussing atmospheric processes. In general, a^+ (often abbreviated to simply a) is used for condensation problems, whereas a^* is commonly preferred for evaporation problems. Values for a^+ can be calculated from Majoube (1971). Although the use of a^+ vs. a^* may simplify calculations, many other people find it more convenient to use the definition of fractionation factor that produces a >1, despite tradition.

As rain condenses, the heavier isotopes of water (mainly $HD^{16}O$ and $H_2{}^{18}O$) are preferentially removed from the air mass (and into the rain), and the air mass consequently becomes progressively lighter in isotopic composition (i.e., higher concentrations of $H_2{}^{16}O$). Hence, the isotopic compositions of successive aliquots of rain become progressively lighter in the heavier isotope due to continuing rainout of the heavy isotopes. This is why the d values of rain become lighter as storms move inland from the ocean. At any point along the storm trajectory (i.e., for some specific fraction f of the total original vapor mass), the$d^{18}O$ of the residual fraction of vapor in the air mass can be calculated by:

$$d \sim d_o + e \ln (f)$$

$$d^{18}O_v \sim d_o{}^{18}O_v + e_{l-v} \cdot \ln f$$

where $d_o{}^{18}O_v$ is the initial d value of the vapor (remember that $\ln x < 0$ for $x < 1$, so that the residual vapor is lighter than the initial vapor). The $d^{18}O$ of the of the rain produced at this point can be determined by:

$$d^{18}O_l \sim d^{18}O_v + e_{l-v}$$

where e_{l-v} (the enrichment of liquid relative to vapor, equivalent to e^+ in the discussion above) is constant. For a system with changing temperature, the relation has to be integrated to account for the change in as a function of temperature.

Another commonly used formulation of the Rayleigh equation for systems with a constant fractionation factor is: $d \sim d_o - e \ln (f)$. In this case, the enrichment factor in the Rayleigh equation has a negative sign, instead of the positive sign shown earlier, because of different definitions for a (and hence for values). The choice of either a^* or its reciprocal value a^+ for the equilibrium fractionation factor is dictated only by convenience; there is no "right" way. If there is any confusion about how the fractionation terms are defined in some paper, just try a few test calculations to make sure the d values for a reaction change in the "right" direction (e.g., with biological reactions, residual reactants get heavier; during condensation of rain, residual vapor gets lighter; etc); if the d values don't change as expected, this probably means that the fractionation factor being used is the inverse of what should be used in the equation.

Evaporation of Water

Evaporation from an open-water surface fractionates the isotopes of hydrogen and oxygen in a manner which depends on a number of environmental parameters, the most important of which is the ambient humidity. This is illustrated for various relative humidities. The higher the humidity, the smaller the change in $d^{18}O$ and dD during evaporation. For example, at 95% humidity, the d values are constant for evaporation of the last 85% of the water. Evaporation results in lines with slopes <8 on a $d^{18}O$ vs. dD plot (i.e., the data plot on lines below the MWL that intersect the MWL at the composition of the original water).

Evaporation at 0% humidity describes open-system evaporation. Note that the two upper diagrams are Rayleigh-type plots, similar but with larger changes in $d^{18}O$ during open-system evaporation. The d values on the curved fractionation lines on the upper diagrams plot along nearly straight lines on the lower $d^{18}O$ vs. D plot. The "length" of the evaporation lines on the $d^{18}O$ vs. dD plot reflect the range of values of water produced during total evaporation under different humidities. For example, the short line for 95% humidity indicates that the water changes little during the entire evaporation process.

Evaporation under almost 100% humidity conditions is more-or-less equivalent to evaporation under closed-system conditions (i.e., isotopic equilibrium is possible), and data for waters plot along a slope of 8 (i.e., along the MWL). However, the shapes of the curves in the upper diagrams for 95% humidity are not the same as for closed-system *equilibrium* fractionation; instead, they are similar to the infinite-reservoir *kinetic* fractionation. Calculated using both an equilibrium fractionation for the phase change and a kinetic fractionation for the diffusion of water vapor across the water-atmosphere interface (Gat and Gonfiantini, 1981). This is also the explanation for the larger open-system fractionations.

The most useful model for the isotope fractionation during evaporation is that of Craig and Gordon (1965). This model assumes that equilibrium conditions apply at the air/water interface (where the humidity is 100%), that there is a constant vertical flux, and that there is no fractionation during fully turbulent transport.

At the water-air interface, there is a balance between two opposing water fluxes: one upward from the water surface and one downward consisting of atmospheric moisture. When the humidity is 100% (i.e., the air is saturated), the upward and downward physical fluxes can become equivalent and their isotopic compositions may then reach equilibrium.

The changes in humidity and corresponding changes in the isotopic composition of vapor across the transition between the water and the free atmosphere are given. Note that where h = 1 (in the so-called "equilibrium vapor" layer between the interface and the boundary layer where the humidity is 100%), the vapor is in equilibrium with the liquid (i.e., $d_v = d - e^*$). When the air is undersaturated (i.e., h < 1), a net evaporative flux is produced. The rate determining step for evaporation is the diffusion of water vapor across the air boundary layer, which occurs in response to the humidity gradient between the surface and the fully turbulent ambient air. The isotopic composition of the evaporated moisture (for either oxygen or hydrogen isotopes) can be formulated as:

$$d_E = (a^* d_w - h d_a - e) / [(1- h) + D/1000] \sim (d_w - h d_a - e) / (1- h)$$

where $e = e^* + D$, $e^* = (1- a^*) \cdot 10^3$, $a^* < 1$, and the variable D is an additional diffusive (kinetic) isotope fractionation which results from the different diffusivities of the water molecules of various isotopic compositions in the liquid-air boundary layer (i.e., an additional fractionation caused by diffusion across the humidity gradient between the "equilibrium vapor layer" and the turbulently mixed vapor sublayer).

Hence, the total fractionation e equals the sum of the equilibrium and kinetic fractionations. d_w and d_a are the isotopic compositions of the surface water and the atmospheric moisture (vapor), respectively, with all parameters in ‰ units. Relative humidity, *h*, is normalized to the saturated vapor pressure at the temperature of the lake surface water, and is written as a fraction < 1. According to the Craig and Gordon (1965) model, De has the form

$$De = (1- h) \cdot q \cdot n \cdot e_k$$

where e_k is a "kinetic" constant with values of 25.1‰ and 28.5‰ for d^2H and $d^{18}O$, respectively, and $0.5 < n < 1$.

The weighting term can be assumed equal to 1 for small bodies of water whose evaporation flux does not perturb the ambient moisture significantly

(Gat, 1995), but has been shown to have a value of 0.88 for the North American Great Lakes (Gat et al., 1994) and a value of about 0.5 for evaporation in the eastern Mediterranean Sea (Gat et al., 1996).

For an open water body, a value of $n = 0.5$ seems appropriate (Gat, 1996). However, for evaporation of water through a stagnant air layer such as in soils (Barnes and Allison, 1988) or leaves (Allison et al., 1985), a value of $n \sim 1$ fits the data reasonably well.

Soils and plants. Note that in some articles the e_k values are modified by multiplication by 1000 (or need to be) because the values for e_k may not be in ‰.

The values of d_E and d_w define a line in $d^{18}O$ vs. dD space called the *evaporation line* whose slope is given by:

$$S = [h(d_a - d_w) + e]_{2H} / [h(d_a - d_w) + e]_{18O}.$$

To preserve mass balance, the initial water composition, the evaporated moisture, and the residual water (such as lake waters or soil waters) must all plot along this same line. The slope of the evaporation line is determined by the air humidity, and the equilibrium and kinetic fractionations (e^* and De), which are dependent themselves on temperature and boundary conditions, respectively. The slopes of evaporation lines range from 3.9 (for humidity = 0) to 6.8 (for humidity = 95%). The d_E value plots above the MWL. This evaporated vapor will mix with ambient vapor d_a to produce vapor with a higher d-excess value than the original vapor, and can affect the d values of later rain from the airmass.

When part of the rained-out moisture is returned to the atmosphere by means of evapo-transpiration, then a simple Rayleigh law no longer applies. The downwind effect of the evapo-transpiration flux on the isotopic composition of the atmospheric moisture and precipitation depends on the details of the evapo-transpiration process.

Transpiration returns precipitated water essentially unfractionated to the atmosphere, despite the complex fractionations in leaf water. Thus, transpiration cancels out the effect of the rainout process. In other words, admixture of transpired waters moves the isotopic composition of the atmospheric moisture back towards more positive d values (i.e., enriched in the heavy isotopic species), as if rain never took place. Under such circumstances, the change in the isotope composition along the air-mass trajectory measures only the net loss of water from the air mass, rather than being a measure of the integrated total rainout. On the other hand, evaporated vapor (d_E) is usually depleted in the heavy isotopic species relative to that of transpired water (d_T) and is actually closer to the composition of the atmospheric moisture. Hence mixing of moisture derived from the evaporation of lake water back into the atmospheric moisture reservoir has a somewhat smaller effect than the addition of transpired water in restoring isotopic composition of the original air mass.

Biological Fractionations

Biological processes are generally unidirectional and are excellent examples of *kinetic* isotope reactions. Organisms preferentially use the lighter isotopic species because of the lower energy "costs" associated with breaking the bonds in these molecules, resulting in significant fractionations between the substrate (heavier) and the biologically mediated product (lighter). Kinetic isotopic fractionations of biologically-mediated processes vary in magnitude, depending on reaction rates, concentrations of products and reactants, environmental conditions, and — in the case of metabolic transformations — species of the organism.

The variability of the fractionations makes interpretation of isotopic data difficult, particularly for nitrogen and sulfur. The fractionations are very different from, and typically larger than, the equivalent equilibrium reaction. The magnitude of the fractionation depends on the reaction pathway utilized (i.e., which is the rate-limiting step) and the relative energies of the bonds severed and formed by the reaction. In general, slower reaction steps show greater isotopic fractionation than faster steps because the organism has time to be more selective (i.e., the organism saves internal energy by preferentially breaking light-isotope bonds).

If the substrate concentration is large enough that the isotopic composition of the reservoir is insignificantly changed by the reaction or if the isotopic ratio of the product is measured within an infinitely short time period (Mariotti et al., 1981), the fractionation factor can be defined. For unidirectional reactions, the change in the isotope ratio of the substrate relative to the fraction of the unreacted substrate can be described by the Rayleigh equation:

$$R_s / R_{so} = f^{(a-1)}$$

where R_s and R_{so} are the ratios of the unreacted and initial substrate, respectively, and f is the fraction of unreacted substrate. The changes in compositions of residual NO_3, incremental N_2 produced, and cumulative N_2 for denitrification with a fractionation factors of = 1.005, 1.010, and 1.020 (i.e., the organism preferentially utilizes the lighter isotope), which are equivalent to a = 0.995, 0.99, and 0.98, respectively. In the final stages of the reaction, when the NO_3 is almost gone, the isotopic compositions of the residual reactant and incremental product increase dramatically, reaching very high values when the reaction is almost complete.

Readers of this book and articles dealing with isotope fractionations must be careful: both fractionation and enrichment factors are defined in various ways by different authors, especially in the biological literature.

Kinetic fractionation factors are typically described in terms of enrichment or discrimination factors, using such symbols as b, e, or D. In particular, the enrichment factor is sometimes defined in reverse (i.e., e_{s-p}), and some researchers define a "discrimination factor" $D_{s/p} = (a_{s/p} - 1)1000$, where *s/p* denotes "substrate relative to products." Good discussions of fractionations associated with biological processes include Hübner (1986) and Fogel and Cifuentes (1993).

A good example of the complexities of kinetic reactions is given by the fractionation between CO_2 and photosynthetic organic carbon. The fractionation can be described by the model (Fogel and Cifuentes, 1993):

$$D = A + (C_i / C_a)(B - A)$$

where D is the isotopic fractionation, *A* is the isotope effect caused by diffusion of CO_2 into the plant (-4.4‰), *B* is the isotope effect caused by enzymatic (photosynthetic) fixation of carbon (-27‰), and C_i / C_a is the ratio of internal to atmospheric CO_2 contents. The magnitude of the fractionation depends on the values of the above parameters. For example, when there is unlimited CO_2 (i.e., $C_i / C_a = 1$), the enzymatic fractionation controls the $d^{13}C$ of the plant, with plant $d^{13}C$ values as low as -36‰ (Fogel and Cifuentes, 1993). Alternatively, if the CO_2 content is limiting ($C_i / C_a << 1$) and the diffusion of CO_2 into the cell is rate determining, $d^{13}C$ values will be strongly affected by the smaller diffusional isotope effect, resulting in more positive $d^{13}C$ values (-20 to -30‰).

3

Stress and Strain

To a scientist, stress is any action or situation that places special physical or psychological demands upon a person, anything that can unbalance his individual equilibrium. And while the physiological response to such demand is surprisingly uniform, the forms of stress are innumerable. Stress may be even but unconscious like the noise of a city or the daily chore of driving the car. Perhaps the one incontestable statement that can be made about stress is that it belongs to everyone- to businessmen and professors, to mother and their children, to factory workers. Stress is a part of fabric of life. Nothing can isolate stress from human beings as is evident from various researches and studies. Stress can be managed but not simply done away with. Today, widely accepted ideas about stress are challenged by new research, and conclusions once firmly established may be turned completely around. The latest evidence suggested. Some stress is necessary to the well being and a lack can be harmful. - Stress definitely causes some serious ailments. -Severe stress makes people accident-prone.

Concept of Stress: Stress is a complex phenomenon. It is very subjective experience. What may be challenge for one will be a stressor for another. It depends largely on background experiences, temperament and environmental conditions. Stress is a part of life and is generated by constantly changing situations that a person must face. The term stress refers to an internal state, which results from frustrating or unsatisfying conditions. A certain level of stress is unavoidable. Because of its complex nature stress has been studied for many years by researchers in psychology, sociology and medicine.

Defining Stress : Defining stress is a very complex matter, which is the subject of different analyses and continuous debate among experts. Beyond the details of this debate, a general consensus can be reached about a definition of stress, which is centered around the idea of a perceived imbalance in the interface between an individual, the environment and other individuals. When people are faced with demands from others or demands from the physical or psycho-social environment to which they feel unable to adequately respond, a reaction of the organism is activated to cope with the situation. The nature of this response depends upon a combination of different elements, including the extent of the demand, the personal characteristics and coping resources of the person, the constraints on the person in trying to cope and the support received from others.)

Stress is involved in an environmental situation that perceived as presenting demand which threatens to exceed the person's capabilities and resources for meeting it, under conditions where he or she expects a substantial differential in the rewards and costs from meeting the demand versus not meeting it.

Stress is the term often used to describe distress, fatigue and feelings of not being able to cope. The term stress has been derived from the Latin word 'stringer' which means to draw tight. The term was used to refer the hardship, strain, adversity or affiction. Stress is an integral part of natural fabric of life. It refers both to the circumstances that place physical or psychological demands on an individual and to the emotional reactions experiences in these situations (Hazards,1994). Although, the adverse effects of stress on physical health and emotional well being are increasingly recognised, there is little agreement among experts on the definition of stress: - According to Selye (1976), stress is caused by physiological, psychological and environmental demands. When confronted with stressors, the body creates extra energy and stress occurs because our bodies do not use up all of the extra energy that has been created. Selye first described this reaction in 1936 and coined it the General Adapt ion Syndrome(GAS). The GAS includes three distinct stages: a) alarm reaction, b) stage of resistance c) stage of exhaustion According to Lazaras, (1976): stress occurs when there are demands on the person, which taxes or exceeds his adjustive resources. According to Spielberger, (1979): the term stress is used to refer to a complex psycho- biological process that consists of three major elements. This process is initiated by a situation or stimulus that is potentially harmful or dangerous stressor. If a stressor is interpreted as dangerous or threatening, an anxiety reaction will be elicited. Thus the definition of stress refers to the following temporal events.

According to Steinberg and Ritzmann, (1990): Stress can be defined as "an under load or overload of matter, energy or information input to, or output from, a living system." According to Levine and Ursin, (1991): "Stress is a part of an adaptive biological system, where a state is created when a central processor registers an informational discrepancy.". According to Humphrey, (1992): In essence, stress can be considered as "any factor, acting internally or externally, that makes it difficult to adapt and that induces increased effort on the part of the person to maintain a state of equilibrium both internally and with the external environment." According to Levi, (1996): "Stress is cost by a multitude of demands (Stressors) such an inadequate fit between what we need and what we capable of, and what our environment offers and what it demands of us." According to Bernik, (1997): "Stress designates the aggression itself leading to discomfort, or the consequences of it. It is our organism's response to a challenge, be it right or wrong." According to Bowman, (1998): "Stress is the body's automatic response to any physical or mental demand placed upon it. When pressures are threatening, the body rushes to supply protection by turning on 'the juices' and preparing to defend itself. It's the 'flight or fight' response in action."

Occupational Stress: Occupational stress can be defined as the harmful physical and emotional responses that occur when the requirements of the job do not match the capabilities, resources, or needs of the worker. Job stress can lead to poor health and even injury. The concept of Occupational stress is often confused with challenge, but these concepts are not the same. Challenge energizes us psychologically and physically, and it motivates us to learn new skills and master our Occupations. When a challenge is met, we feel relaxed and satisfied (NIOSH,1999).

Thus, challenge is an important ingredient for healthy and productive work. The importance of challenge in our work lives is probably what people are referring to when they say, "a little bit of stress is good for you. Occupational stress is that which derives specifically from conditions in the work place. These may either cause stress initially or aggravate the stress already present from other sources. In today's typical workplace, stress is seen as becoming increasingly more common. People appear to be working longer hours, taking on higher level of responsibilities and exerting themselves even more strenuously to meet rising expectations about Occupational performance. Competition is sharp. There is always someone else ready to "step into one's shoes" should one be found wanting.

Definitions of Occupational Stress: According to Kyriacou (1987), defines "teacher stress as the experience by a teacher of unpleasant emotions such as tension, frustration, anger and depression resulting from aspects of his work as a teacher."

According to Okebukola and Jegede (1989), defined occupational stress as "a condition of mental and physical exertion brought about as a result of harassing events or dissatisfying elements or general features of the working environment." According to Borg (1990), conceptualizes teacher stress as a negative and potentially harmful to teachers' health. The key element in the definition is the teacher's perception of threat based on the following three aspects of his job circumstances. 1. that demands are being made on him. 2. that he is unable to meet or has difficulty in meeting these demands. 3. that failure to meet these demands threatens his mental/physical well being." According to United States National Institute of Occupational Safety and Health, Cincinnati, (1999), Job stress can be defined as "the harmful physical and emotional responses the occur when the requirements of the job do not match the capabilities, resources, or needs of the worker. Job stress can lead to poor health and even injury. According to a discussion document presented by United Kingdom Health and Safety Commission, London, (1999), "Stress is the reaction people have to, excessive pressures or other types of demand placed on them. According to Denise Allen, (2002): "Stress is a feeling we experience, when we loose confidence in our capability to cope with a situation. According to European Commission, Directorate General for Employment and Social Affairs, (2005) "The emotional cognitive, behavioral and physiological reaction to aversive and noxious aspects of work, work environments and work organizations. It is a state characterized by high levels of arousal and distress and often by feelings of not coping."

Niosh Approach to Occupational Stress: On the basis of experience and research, NIOSH favors the view that working conditions play a primary role in causing Occupational stress. However, the role of individual factors is not ignored. According to the NIOSH view, exposure to stressful working conditions (called Occupational stressors) can have a direct influence on worker safety and health. But as shown below, individual and other situational factors can intervene to strengthen or weaken this influence. Theresa's need to care for her ill mother is an increasingly common example of an individual or situational factor that may intensify the effects of stressful working conditions. Examples of individual and situational factors that can help to reduce the effects of stressful working conditions include the following:

- Balance between work and family or personal life
- A support network of friends and coworkers
- A relaxed and positive outlook.

CAUSES OF OCCUPATIONAL STRESS

Nearly everyone agrees that Occupational stress results from the interaction of the worker and the conditions of work. Views differ, however, on the importance of worker characteristics versus working conditions as the primary cause of Occupational stress. These differing viewpoints are important because they suggest different ways to prevent stress at work. According to one school of thought, differences in individual characteristics such as personality and coping style are most important in predicting whether certain Occupational conditions will result in stress-in other words, what is stressful for one person may not be a problem for someone else. This viewpoint leads to prevention strategies that focus on workers and ways to help them cope with demanding Occupational conditions. Although the importance of individual differences cannot be ignored, scientific evidence suggests that certain working conditions are stressful to most people. The excessive workload demands and conflicting expectations. Such evidence argues for a greater emphasis on working conditions as the key source of Occupational stress, and for Occupational redesign as a primary prevention strategy.

Work Related Stress; Job stress has been associated with poor mental and physical health. A useful model for understanding how the work environment effects individual health and well-being is provided by (Levi,1996). In this model there are the following components:

1. Stressors: These are aspects of the working environment that cause stress for the individual..
2. Appraisal: The way a stressor is appraised will vary between individuals depending on such things as personality, customs and attitudes.
3. Stress: Stress is produced when the stressor interacts with the individual's appraisal of it to induce emotional, behavioral and physiological reactions. Emotional reactions include anxiety, depression, restlessness and fatigue. Behavioral reactions include increased smoking, overindulgence in food or drink and taking unnecessary risks. Physiological reactions include increased blood pressure, increased or irregular heartbeat, muscular tension and associated pain and heartburn.
4. Disease: The above reactions may result in suffering, illness and death (e.g. through suicide, diseases of the heart and blood vessels, or cancer). This sequence of events may be promoted or counteracted by interacting variables such as coping repertoire, social support, physical environment and nutrition.

Symptoms of Work-related Stress : Defining a clear link between occupational causes, and the resulting symptoms is much harder for a condition such as stress than is it for a disease such as mesothelioma (which is only caused by exposure to asbestos). Because many of the symptoms of stress are generalised - such as increased anxiety, or irritability - it is easy for them to be ascribed to a characteristic of the worker, rather than to a condition of the work. As we will show, however, there is mounting scientific and medical evidence that certain types of work and work organisation do have a measurable, and verifiable impact on the health of workers. The range of symptoms includes the following:

Physical Symptoms Mental Health Symptoms Psychological Symptoms Asthma Irritability Smoking Ulcers Depression Heavy drinking Heart disease

Anxiety Eating Disorders Diabetes Burn out Increased sickness Thyroid disorders Withdrawal Low self esteem

Some degree of stress is a normal part of life and provides part of the stimulus to learn and grow, without having an adverse effect on health. When stress is intense, continuous or repeated - as is often the case with occupational stress - ill health can result (Hazards,1994). The experience of stress can affect the way individuals think, feel and behave, and can also cause physiological changes. Many of the short and long term illnesses caused by stress can be accounted for by the physiological changes that take place when the body is placed under stress. From the documented evidence, it is clear that as far as work life is concerned extreme stress is so aversive to employees that they will try to avoid it by withdrawing psychologically (through disinterest or lack of involvement in the occupation etc.). Excessive stress can destroy the quality of life and also effect family life. Workers under stress are far more likely to have accidents than workers in low stress jobs, and are much more likely to have to take time off work for stress-related sickness. In jobs where work overload is the cause of the stress, the workers find that they have to take time off to deal with the stress, only to return to work to find that the already unmanageable workload has substantially increased in their absence, thereby increasing the source of the stress and fuelling a vicious cycle which may ultimately lead to a complete breakdown in health (Selye,1976).

Significance of Study : Occupational stress can be inadvertently linked to success or failure at one's job. The general impression about occupational stress is the feeling of failure due to work overload. But if this is the case and so simple a problem then merely by reducing the amount of work, occupational stress could have been done away with. However the problem is not that easy to pinpoint. It is here that a comparative investigation of the reasons of stress in different occupations becomes important. Herein lies the most crucial significance of the study. To combat a problem the awareness of the conditions, which lead to it, are very important. Stress is a part of everyone's daily life. It means that the person cannot cope with the demands put forward by his or her work, which is opposite to their expectations of rewards and success. It affects both the person concerned and the relationships he or she forms in the society be it with family or friends. Although the importance of individual differences cannot be ignored, scientific evidence suggests that certain working conditions are stressful to most people. The excessive workload demands and conflicting expectations and puts a greater emphasis on working conditions as the key source of job stress, and for job redesign as a primary prevention strategy. In jobs where work overload is the cause of the stress, the workers find that they have to take time off to deal with the stress, only to return to work to find that the already unmanageable workload has substantially increased in their absence, thereby increasing the source of the stress and fuelling a vicious cycle which may ultimately lead to a complete breakdown in health. At times the work stress becomes so extreme that the workers grow aversive of it and they try to avoid it by withdrawing either psychologically (through disinterest or lack of involvement in the job etc.). or physically through absenteeism, frequently reporting late for work and even while working an attitude of lethargy persists. In this present era of cutthroat competition the idea of being perfect becomes very necessary to strive and become successful. The worker has to be perfect in his job or else he will be replaced or at least lag behind in his work leading to stress. In India the problem of stress management is gaining more and more importance due to the new privatized nature of the economy. People are leaving behind the cozy atmosphere of government jobs and joining the private sector where there is no end to the amount of work that a person can undertake. In this environment coping with stress becomes very important. One has to be aware of the problem well in advance to be able to deal with it. The study becomes very important to be aware of the problems of the present, then build strategies for the future, and also consider the problems that may arise. Stress factor of males and females according to the age of the worker and the kind of work that he performs are key areas to identify the problems.

Objectives of the Study: To study occupational stress among employees of different careers of Chandigarh. 2. To study occupational stress among males and females employees of different careers of Chandigarh. 3. To study occupational stress among less and more experienced employees of different careers of Chandigarh.

There will be significant difference of occupational stress among employees of different careers of Chandigarh. 2. There will be significant difference of occupational stress among males and females employees of different careers of Chandigarh. 3. There will be significant difference of occupational stress among less and more experienced employees of different careers of Chandigarh.

The sample size will be 200 employees. 2. The area of study will be limited to Chandigarh. 3. For data collection, occupational stress inventory will only be used.

Theoretical background: Continuum mechanics deals with deformable bodies, as opposed to rigid bodies. The stresses considered in continuum mechanics are only those produced by deformation of the body, *sc.* relative changes are considered rather than absolute values. A body is considered stress-free if the only forces present are those inter-atomic forces (ionic, metallic, and van der Waals forces) required to hold the body together and to keep its shape in the absence of all external influences, including gravitational attraction. Stresses generated during manufacture of the body to a specific configuration are also excluded.

Following classical Newtonian and Eulerian dynamics, the motion of a material body is produced by the action of externally applied forces which are assumed to be of two kinds: surface forces and body forces.

Surface forces, or contact forces, can act either on the bounding surface of the body, as a result of mechanical contact with other bodies, or on imaginary internal surfaces that bind portions of the body, as a result of the mechanical interaction between the parts of the body to either side of the surface (#Euler-Cauchy's stress principle). When external contact forces act on a body, internal contact forces pass from point to point inside the body to balance their action, according to Newton's second law of motion of conservation of linear momentum and angular momentum. These laws are called Euler's equations of motion for continuous bodies. The internal contact forces are related to the body's deformation through constitutive equations. This article provides mathematical descriptions of internal contact forces and how they relate to the body's motion, independent of the body's material makeup.

Stress can be thought as a measure of the internal contact forces' intensity acting between particles of the body across imaginary internal surfaces. In other words, stress is a measure of the average quantity of force exerted per unit area of the surface on which these internal forces act. The intensity of contact forces is in inverse proportion to the contact area. For example, if a force applied to a small area is compared to a distributed load of the same resultant magnitude applied to a larger area, one finds that the effects or intensities of these two forces are locally different because the stresses are not the same.

Body forces originate from sources outside of the body that act on its volume (or mass). This implies that the *internal forces* manifest through the contact forces alone. These forces arise from the presence of the body in force fields, (*e.g.*, a gravitational field). As the mass of a continuous body is assumed to be continuously distributed, any force originating from the mass is also continuously distributed. Thus, body forces are assumed to be continuous over the body's volume.

The density of internal forces at every point in a deformable body is not necessarily even, *i.e.* there is a distribution of stresses. This variation of internal forces is governed by the laws of conservation of linear and angular momentum, which normally apply to a mass particle but extend in continuum mechanics to a body of continuously distributed mass. If a body is represented as an assemblage of discrete particles, each governed by Newton's laws of motion, then Euler's equations can be derived from Newton's laws. Euler's equations can, however, be taken as axioms describing the laws of motion for extended bodies, independently of any particle structure.

Rocks Deform

Many students have a difficult time realizing that rocks can bend or break. They also may have difficulty imagining the forces necessary to fold or fault rocks or comprehending that the seemingly constant Earth can change dramatically over time. This is especially true of students who live in tectonically stable areas. If students are to understand the basics of stress and strain, they must overcome this barrier since it will be difficult to examine the causes and conditions of deformation if students cannot comprehend deformation. It is often helpful to have students create analog models of the structures present in rock photos or hand samples.

In order to show students that rocks deform, pictures and hand samples of *real* faulted and folded rocks at a variety of scales can be used. There are several good collections of these types of images such as the AGI Earthscience World Image Bank, Martin Miller's collection, or the National Geophysical Data Center Faults slide set.

Stress causes strain, strain results in structures: Many geologists consider it important for introductory students to understand that visible structures are a record of the stress and physical conditions in the Earth. As a result, the differences between stress, strain and structures formed during strain become key concepts.

Stress, strain and structure start with the same three letters, yet mean very different things. These words are also used differently in geology than in common usage in English, which can cause confusion. However, here are some tricks that I use to remember:

- Stress is the same as pressure. When you are under pressure, you are stressed!
- Stress can happen with out strain, but strain cannot happen without stress.

Look at this rock I am squeezing in my hand.

- Is it stressed? (Yes, it is under pressure.)
- Is it strained? (No, it hasn't changed shape.)

Now look at this rock with a fold in it.

- Is it under stress? (No, it is not under pressure).
- Is it straining? (No, it isn't currently changing shape.)
- Does it have structure? (Yes, there is a fold.)

In addition to Silly Putty® and Play-Doh®, wooden blocks with layers drawn in or a compression/squeeze tank filled with layered sand or breakfast cereals also model structural features well. Analogs, however, are difficult to scale appropriately (both in time and space) to the gigantic scale on which geologic structures form. Students may still have difficulty understanding the tremendous scale of forces needed to bend or break rock and the long time scales involved to generate structures. Make sure that you make it clear to your students that these pitfalls exist. More detailed ideas for analogs are available at Teaching Structural Geology analog materials web page.

Once students have mastered the connections among stress, strain and structure, I develop a 3 x 2 table of different structures that form under differing stress and strain conditions. Let's look at what features are found under different stress conditions and with different styles of strain. We'll do this by making a table. What are the three types of stress? Compression, tension, and shearing. Now, what are the 2 types of permanent deformation? Ductile and brittle. Let's make a table that is three columns by two rows and fill it in with appropriate structures! When we are finished, we should have 6 kinds of deformation features.

Different conditions lead to different deformation styles: There are many factors that contribute to the style of the deformation in a rock, including pressure, temperature, rock composition, presence or absence of fluids, type of stress, rate of stress, and others. However, the type of stress, the rate of stress and the temperature may be the most critical factors for most introductory students.

Silly Putty® is a material, just like rocks, that can deform either plastically or brittlely. What controls how it will deform?

- Temperature: Cold Silly Putty is easy to break, but warm Silly Putty is very plastic.
- Strain rate: If I pull it apart quickly it breaks, but if I pull it slowly, it stretches (deforms plastically).
- Type of stress: Finally, pick a strong student, and have him or her try to break the silly putty using compressive stress. As you can see, this is almost impossible. Now have a student try to break it using tension. This is much easier. Most materials are more easily broken (or otherwise deformed) in tension than in compression; we say that they are weaker in tension, or stronger in compression.

Temperature, strain rate, and type of stress are also key factors in deformation within glaciers. This provides a an opportunity to revisit these concepts later in the term.

Relating faults to stress - hanging walls, footwalls, and different types of faults: One of the goals of structural geology is to relate the nature of deformation to the stress that caused it. Therefore, it is important that students be able to distinguish between normal faults (generated by tension) and reverse faults (generated by compression).

Wooden blocks are a valuable tool for teaching about normal and reverse faults. Using three blocks cut on an angle, horsts and grabens can be generated. Pull the blocks apart to create a graben; push them together to make a horst. The advantage of using 3 blocks is that students can see that it is not the orientation of the fault that matters, but the movement on the fault. Because they can see whether I am extending or compressing the blocks, they develop an intuitive sense of the difference between normal and reverse faults. However, students typically still need to learn the difference between the hanging wall and footwall of a fault to be able to accurately determine whether a fault is normal or reverse and what kind of stress caused it.

Faults are places where rocks have been broken and offset. It is not uncommon for fluids to have flowed along the break during deformation, leaving valuable minerals along the fault. As a result, many mines are constructed along fault surfaces. Because of this, one side of the fault is called the hanging wall (the surface from which a miner's lantern would be hung) and one side is called the footwall (the surface on which the miner would walk.)

Here's another way to think of it: the hanging wall block is always above the fault plane, while the foot wall block is always below the fault plane. To see this, put a point on the fault and draw a vertical arrow pointing up. This arrow points into the hanging wall. An arrow pointing straight down points into the footwall. Take a look at the slide that shows the fault and arrows indicating movement. Some students think the footwall looks like a foot. See how the hanging wall is resting, or hanging, on the footwall?

Once students understand the difference between a hanging wall and a footwall, most of them have little trouble remembering that in a reverse fault the hanging wall moves up, indicating compression, and in a normal fault the hanging wall moves down, indicating extension.

As your students can see from these block models, horizontal forces can cause rocks to move along faults that are at an angle to the rock layering. Given that idea, your students can use some basic trigonometric functions to examine the relationship between horizontal strain (the amount of stretching or shortening in a horizontal direction) and displacement on a fault surface (the amount of movement on the fault itself). Since this relationship is dependent on the angle of the fault from horizontal, the angle of the fault is a critical component of how faults accommodate shortening or extension. The Seattle fault is a large reverse fault that cuts across and underneath the Seattle, Washington, metropolitan area and its nearly 2 million residents. The Seattle fault accommodates about 1 millimeter of shortening per year. However, because the fault plane itself is not well exposed and or discernable in seismic profiles, we don't know what angle this fault makes to horizontal. If the fault is low-angle, at about 25 degrees from horizontal, then to accommodate the 1 mm of shortening it must move 1.1 mm/yr on average. If, however, it is at a more steep 60 degrees, it would need to move 2 mm/yr on average. Since the displacement on the fault is a primary factor in determining the magnitude of an earthquake, the Seattle fault would need to move either twice as often or have much larger quakes if it is at a steep angle.

Relating analogs to the real Earth: We often use analogies and analog materials (Silly Putty, sand, wooden blocks, etc.) to illustrate the concepts of stress, strain and the deformation of rocks. However, students sometimes have difficulty relating these materials and their behaviors to the Earth and real rocks. For these students, it may be useful to discuss the rates and magnitudes of deformation in the Earth and the differences between rocks and analog materials. For instance, the rocks at plate boundaries often experience a few centimeters of deformation in a year, but the forces on them are enough to move continents. The size and slowness of these processes are an important concept to communicate, even if they are on a scale that is almost impossible to comprehend. I sometimes tell students that their fingernails are growing at about the same rate the plates are moving, to help them overcome this difficulty.

STRAIN MARKERS IN NATURALLY DEFORMED ROCKS

Strain analysis is a subdiscipline of structural geology concerned with the measurement of the shape changes (strains) associated with geological deformation. Stress analysis (or palaeostress analysis) is a closely allied field devoted to the estimation of past stresses (palaeostresses) from observed rock strains.

The calculation of geological strain involves the comparison of the original and deformed shapes of objects (*strain markers*) enclosed within the rock. Examples of strain markers are fossils and the outlines of sedimentary particles. Most useful are objects with known original dimensions. For example, the original length of certain graptolites can be estimated by counting the thecae present.

A measure of length change is the *stretch*; the ratio of new to old lengths. Such measurements commonly reveal that the value of stretch varies according to the direction of the marker being measured. This accords with the concept of the *strain ellipsoid*; a representation of the state of strain, which considers the final form adopted by an original sphere with unit radius. The results of strain analysis are usually expressed in terms of this ellipsoid, which can be completely described by six quantities: the stretches in the directions of the principal axes ($S_1 = S_2 = S_3$ (and three angles defining the orientation of those axes. Ooids, that is grains in limestones with near-spherical original shapes, are examples of strain markers which allow direct estimation of strain ellipsoid parameters, though even here the usual lack of information on sizes means that only ratios of the principal stretches (R_s) can be determined. Only where *principal stretches* are known can volumetric strain be assessed. A large number of different techniques, many of ingenious design, have been developed for the treatment of natural strain gauges.

Naturally Deformed Quartz-rich Rocks

In naturally deformed quartz-rich rocks, the microstructures show systematic differences with increasing temperatures of deformation, similar to the different dislocation creep regimes in experimentally deformed examples. The strain rate and stress conditions of the naturally deformed rocks usually are unknown or poorly constrained, so that a direct correlation with dislocation creep regimes could be difficult. However, the dominant recrystallization mechanisms can be determined from the microstructures and can be compared with the experimentally established dislocation creep regimes. Three microstructural regimes corresponding to three main mechanisms of dynamic recrystallization can be distinguished:

1. Bulging recrystallization which is dominated by local grain boundary migration (slow migration) and occurs at the lowest temperatures of deformation. The grain boundary lobes are very small. Favorite sites for bulging are triple junctions, and - if present - fractures.
2. Progressive subgrain rotation which is dominated by polygonization of old grains and formation of newly recrystallized grains. This recrystallization occurs at intermediate temperatures.
3. Grain boundary migration recrystallization which is dominated by fast grain boundary migration and occurs at high temperatures. During this recrystallization, whole grains may be swept. Progressive subgrain rotation is only important for the initial formation of new grains.

Generally, the recrystallization mechanisms listed above as 1,2,3 correspond approximately to the dominant recrystallization processes identified in the respective experimental dislocation creep regimes 1,2,3.

The Heavitree quartzite is from the Ruby Gap duplex (RGD) which forms a part of the internal ductile zones of the Alice Springs orogen in central Australia. The RGD consists of 5 thrust sheets of Heavitree quartzite deformed under greenschist-facies conditions (less than 400°C). Sheets 1, 2 and 3 form an imbricate system. Finite strain and temperature of deformation generally increase upward through the duplex. The microstructures are typical of the full spectrum of dislocation creep regimes.

The chert samples are from the Warrawoona syncline which is part of an Archean greenstone belt accumulated between 3450 Ma and 3320 Ma. The syncline is a tight keel structure developed between two granitic domes. In the axis of the syncline, a chert series was deformed and recrystallized under greenschist-facies conditions. The tectonites are very strongly lineated in the central part of the syncline, and shape and crystallographic fabric analyses indicate a deformation in constriction. During deformation-recrystallization, the chert underwent substantial grain growth.

The quartz veins are from the Tonale Line, a major strike slip fault in the Alps. At the eastern end, the synkinematic Adamello intrusion has imposed a thermal gradient across the fault. The quartz samples shown here come from the Edolo shists which were deformed under greenschist-facies conditions.

Naturally Deformed Feldspar Rocks

In naturally deformed rocks, dislocation creep in feldspar is usually found above temperatures of approximately 500°C. Below these temperatures, deformation takes place by fracturing. However, this simple temperature-classification of deformation mechanisms in feldspar is complicated by the dependence of feldspar chemistry on ambient P,T-conditions and bulk rock composition. Compositional disequilibrium may be a driving force that leads to recrystallization of plagioclase under much lower temperatures than 500°C in deformed rocks. In such cases, recrystallization is syndeformational but not dynamic, because it is not driven by strain energy alone and is better termed neo-mineralisation. In order to determine the recrystallization mechanism in plagioclase one needs to compare the chemical composition of recrystallized and host grains; if the compositions are identical the process was dynamic recrystallization, but if they are different it was neo-mineralization.

Dynamic recrystallization by regime 1 dislocation creep, with no chemical change, is very rare in nature. Samples shown here are from the Corvatsch shear zone (Engadine, Swiss Alps). Examples of low temperature recrystallization (low to mid-greenschist facies) are shown in micrographs 23 and 24. Fracturing is the most important contribution to the deformation of plagioclase in these examples, accompanied by neo-mineralization.

With the exception of photo 47, the series of samples of regime 2 come from a shear zone between Sogndal and Kaupanger (Norway), in an anorthosite body of the Jotun nappe, which is part of the "middle allochthonous unit" of the Caledonides. The samples represent stages of progressively higher shear strain along a strain gradient from less deformed boudins into strongly deformed mylonites. Deformation took place around 700°C at pressures below

900 MPa. The chemical composition of the recrystallized plagioclase grains is the same as that of the porphyroclasts, so that the formation of new grains represents only strain-induced dynamic recrystallization. Dynamic recrystallization took place dominantly by progressive subgrain rotation with some local grain boundary bulging (largely regime 2 dislocation creep).

STRESS AND STRAIN - ROCK DEFORMATION

Stress - Pressure Applied to Rock: Rock can be subject to several different kinds of stress:

lithostatic stress: Rock beneath the Earth's surface experiences equal pressure exerted on it from all directions because of the weight of the overlying rock.

It is like the hydrostatic stress (water pressure) that a person feels pressing all around their body when diving down deep in water.

Differential (deviatoric) stress: In many cases, rock may experience an additional, unequal stress due to tectonic forces. There are three basic kinds.

Tensional stress (stretching)

Compressional stress (squeezing)

Shearing stress (side to side shearing).

Stress Analysis

Stress analysis is an engineering (e.g., civil engineering and mechanical engineering) discipline that determines the stress in materials and structures subjected to static or dynamic forces or loads. A stress analysis is required for the study and design of structures, e.g., tunnels, dams, mechanical parts, and structural frames among others, under prescribed or expected loads. Stress analysis may be applied as a design step to structures that do not yet exist.

The aim of the analysis is usually to determine whether the element or collection of elements, usually referred to as a structure, can safely withstand the specified forces. This is achieved when the determined stress from the applied force(s) is less than the ultimate tensile strength, ultimate compressive strength or fatigue strength the material is known to be able to withstand, though ordinarily a factor of safety is applied in design.

Analysis may be performed through mathematical modelling or simulation, through experimental testing procedures, or a combination of techniques.

Engineering quantities are usually measured in megapascals (MPa) or gigapascals (GPa). In imperial units, stress is expressed in pounds-force per square inch (psi) or kilopounds-force per square inch (ksi).

Analysis methods: The analysis of stress within a body implies the determination at each point of the body of the magnitudes of the nine stress components. In other words, it is the determination of the internal distribution of stresses.

A key part of analysis involves determining the type of loads acting on a structure, including tension, compression, shear, torsion, bending, or combinations of such loads.

When forces are applied, or expected to be applied, repeatedly, nearly all materials will rupture or fail at a lower stress than they would otherwise. The analysis to determine stresses under these cyclic loading conditions is termed fatigue analysis and is most often applied to aerodynamic structural systems.

Modelling: To determine the distribution of stress in a structure it is necessary to solve a boundary-value problem by specifying the boundary conditions, i.e. displacements and/or forces on the boundary. Constitutive equations, such as e.g. Hooke's Law for linear elastic materials, are used to describe the stress:strain relationship in these calculations. A boundary-value problem based on the theory of elasticity is applied to structures expected to deform elastically, i.e. infinitesimal strains, under design loads. When the loads applied to the structure induce plastic deformations, the theory of plasticity is implemented.

Approximate solutions for boundary-value problems can be obtained through the use of numerical methods such as the finite element method, the finite difference method, and the boundary element method, which are implemented in computer programs. Analytical or close-form solutions can be obtained for simple geometries, constitutive relations, and boundary conditions.

All real objects occupy a three-dimensional space. The stress analysis can be simplified in cases where the physical dimensions and the loading conditions allows the structure to be assumed as one-dimensional or two-dimensional. For a two-dimensional analysis a plane stress or a plane strain condition can be assumed.

Experimental testing: Stress analysis can be performed experimentally by applying forces to a test element or structure and then determining the resulting stress using sensors. In this case the process would more properly be known as *testing* (destructive or non-destructive). Experimental methods may be used in cases where mathematical approaches are cumbersome or inaccurate. Special equipment appropriate to the experimental method is used to apply the static or dynamic loading.

There are a number of experimental methods which may be used:

Tensile testing is a fundamental materials science test in which a sample is subjected to uniaxial tension until failure. The results from the test are commonly used to select a material for an application, for quality control, and to predict how a material will react under other types of forces. Properties that are directly measured via a tensile test are ultimate tensile strength, maximum elongation and reduction in area. From these measurements properties such as Young's modulus, Poisson's ratio, yield strength, and strain-hardening characteristics can be determined.

The photoelastic method relies on the physical phenomenon of birefringence. Unlike the analytical methods of stress determination, photoelasticity gives a fairly accurate picture of stress distribution even around abrupt discontinuities in a material. The method serves as an important tool for determining the critical stress points in a material and is often used for determining stress concentration factors in irregular geometries. Birefringence is exhibited by certain transparent materials. A ray of light passing through a birefringent material experiences two refractive indices. This double refraction is exhibited by many optical crystals. But photoelastic materials exhibit the property of birefringence only on the application of stress, and the magnitude of the refractive indices at each point in the material is directly related to the state of stress at that point. A model component is created made of photoelastic material with similar geometry to that of the structure on which stress analysis is to be performed. This ensures that the state of the stress in the model is similar to the state of the stress in the structure.

Dynamic mechanical analysis is a technique used to study and characterize viscoelastic materials, particularly polymers. Polymers composed of long molecular chains have unique viscoelastic properties, which combine the characteristics of elastic solids and Newtonian fluids. The viscoelastic property of polymer is studied by dynamic mechanical analysis where a sinusoidal force (stress) is applied to a material and the resulting displacement (strain) is measured. For a perfectly elastic solid, the resulting strain and the stress will be perfectly in phase. For a purely viscous fluid, there will be a 90 degree phase lag of strain with respect to stress. Viscoelastic polymers have the characteristics in between where some phase lag will occur during DMA tests. Analyzers are made for both stress and strain control. In strain control, a probe is displaced and the resulting stress of the sample is measured. In stress control, a set force is applied and several other experimental conditions (temperature, frequency, or time) can be varied.

Factor of safety: The factor of safety is a design requirement for the structure based on the uncertainty in loads, material strength, and consequences of failure. In design of structures, calculated stresses are restricted to be less than an specified allowable stress, also known as working or designed stress, that is chosen as some fraction of the yield strength or of the ultimate strength of the material which the structure is made of. The ratio of the ultimate stress to the allowable stress is defined as the factor of safety.

Laboratory test are usually performed on material samples in order to determine the yield strength and the ultimate strength that the material can withstand before failure. Often a separate factor of safety is applied to the yield strength and to the ultimate strength. The factor of safety on yield strength is to prevent detrimental deformations and the factor of safety on ultimate strength is to prevent collapse.

Load transfer: The evaluation of loads and stresses within structures is directed to finding the load transfer path. Loads will be transferred by physical contact between the various component parts and within structures. The load transfer may be identified visually, or by simple logic for simple structures. For more complex structures, more complex methods such as theoretical solid mechanics or by numerical methods may be required. Numerical methods include direct stiffness method which is also referred to as the finite element method.

The object is to determine the critical stresses in each part, and compare them to the strength of the material.

For parts that have broken in service, a forensic engineering or failure analysis is performed to identify weakness, where broken parts are analysed for the cause or causes of failure. The method seeks to identify the weakest component in the load path. If this is the part which actually failed, then it may corroborate independent evidence of the failure. If not, then another explanation has to be sought, such as a defective part with a lower tensile strength than it should for example.

Uniaxial stress: If two of the dimensions of the object are very large or very small compared to the others, the object may be modelled as one-dimensional. In this case the stress tensor has only one component and is indistinguishable from a scalar. One-dimensional objects include a piece of wire loaded at the ends and a metal sheet loaded on the face and viewed up close and through the cross section.

When a structural element is subjected to tension or compression its length will tend to elongate or shorten, and its cross-sectional area changes by an amount that depends on the Poisson's ratio of the material. In engineering applications, structural members experience small deformations and the reduction in cross-sectional area is very small and can be neglected, i.e., the cross-sectional area is assumed constant during deformation. For this case, the stress is called *engineering stress* or *nominal stress*. In some other cases, e.g., elastomers and plastic materials, the change in cross-sectional area is significant, and the stress must be calculated assuming the current cross-sectional area instead of the initial cross-sectional area.

Strain - Rock Deformation in Response to Stress

Rock responds to stress differently depending on the pressure and temperature (depth in Earth) and mineralogic composition of the rock.

Elastic deformation: For small differential stresses, less than the*yield strength,* rock deforms like a spring. It changes shape by a very small amount in response to the stress, but the deformation is not permanent. If the stress could be reversed the rock would return to its original shape.

Brittle deformation: Near the Earth's surface rock behaves in its familiar brittle fashion. If a differential stress is applied that is greater than the rock's yield strength, the rock fractures. *It breaks.* Note: the part of the rock that didn't break springs back to its original shape. This *elastic rebound* is what causes earthquakes.

Ductile deformation: Deeper than 10-20 km the enormous lithostatic stress makes it nearly impossible to produce a fracture (crack - with space between masses of rock) but the high temperature makes rock softer, less brittle, more malleable. Rock undergoes plastic deformation when a differential stress is applied that is stronger than its yield strength. *It flows.* This occurs in the lower continental crust and in the mantle

There are three mechanisms by which ductile deformation occurs depending principally on the temperature.

Plastic flow occurs at the relatively low temperatures in the lower crust and uppermost mantle. Rock deforms by gradual creep along crystal grain boundaries and planes within crystal lattices. This method leads to *work hardening* because crossing planes and grain boundaries soon lock the creep planes.

Power law creep occurs where the temperature is greater than 55% of the melting temperature for that depth (pressure). Movement occurs along crystallographic planes and grain boundaries, as in plastic flow, but diffusion of ions/atoms at the higher temperatures allows for continuous and large deformation. High stresses at locked points cause the atoms to diffuse to lower stress points. This is called *dynamic recrystallization.*

Diffusion creep occurs where the temperature is greater than 85% of the melting temperature for that depth (pressure). Deformation/flow is largely accomplished by ionic/atomic diffusion and continuous recrystallization.

Stress and Strain Analysis and Measurement

The engineering design of structures using polymers requires a thorough knowledge of the basic principles of stress and strain analysis and measurement.

Readers of this book should have a fundamental knowledge of stress and strain from a course in elementary solid mechanics and from an introductory course in materials. Therefore, we do not rigorously derive from first principles all the necessary concepts. However, in this chapter we provide a review of the fundamentals and lay out consistent notation used in the remainder of the text. It should be emphasized that the interpretations of stress and strain distributions in polymers and the properties derived from the standpoint of the traditional analysis given in this chapter are approximate and not applicable to viscoelastic polymers under all circumstances.

Some Important and Useful Definitions: In elementary mechanics of materials (Strength of Materials or the first undergraduate course in solid mechanics) as well as in an introductory graduate elasticity course five fundamental assumptions are normally made about the characteristics of the materials for which the analysis is valid. These assumptions require the material to be,

- Linear
- Homogeneous
- Isotropic
- Elastic
- Continuum.

Provided that a material has these characteristics, be it a metal or polymer, the elementary stress analysis of bars, beams, frames, pressure vessels, columns, etc. using these assumptions is quite accurate and useful.

However, when these assumptions are violated serious errors can occur if the same analysis approaches are used. It is therefore incumbent upon engineers to thoroughly understand these fundamental definitions as well as how to determine if they are appropriate for a given situation. As a result, the reader is encouraged to gain a thorough understanding of the following terms:

Linearity: Two types of linearity are normally assumed: Material linearity (Hookean stress-strain behavior) or linear relation between stress and strain; Geometric linearity or small strains and deformation. Elastic: Deformations due to external loads are completely and instantaneously reversible upon load removal.

Continuum: Matter is continuously distributed for all size scales, i.e. there are no holes or voids.

Homogeneous: Material properties are the same at every point or material properties are invariant upon translation.

Inhomogeneous or Heterogeneous: Material properties are not the same at every point or material properties vary upon translation.

Amorphous: Chaotic or having structure without order. An example would be glass or most metals on a macroscopic scale.

Crystalline: Having order or a regular structural arrangement. An example would be naturally occurring crystals such as salt or many metals on the microscopic scale within grain boundaries.

Isotropic: Materials which have the same mechanical properties in all directions at an arbitrary point or materials whose properties are invariant upon rotation of axes at a point. Amorphous materials are isotropic.

Anisotropic: Materials which have mechanical properties which are not the same in different directions at a point or materials whose properties vary with rotation at a point. Crystalline materials are anisotropic.

Plastic: The word comes from the Latin word plasticus, and from the Greek words plastikos which in turn is derived from plastos (meaning molded) and from plassein (meaning to mold). Unfortunately, this term is often used as a generic name for a polymer probably because many of the early polymers (cellulose, polyesters, etc.) appear to yield and/or flow in a similar manner to metals and could be easily molded. However, not all polymers are moldable, exhibit plastic flow or a definitive yield point.

Viscoelasticity or Rheology: The study of materials whose mechanical properties have characteristics of both solid and fluid materials. Viscoelasticity is a term often used by those whose primary interest is solid mechanics while rheology is a term often used by those whose primary interest is fluid mechanics. The term also implies that mechanical properties are a function of time due to the intrinsic nature of a material and that the material possesses a memory (fading) of past events. The latter separates such materials from those with time dependent properties due primarily to changing environments or corrosion. All polymers (fluid or solid) have time or temperature domains in which they are viscoelastic. Polymer: The word Polymer originates from the Greek word "polymeros" which means many-membered, (Clegg and Collyer 1993). Often the word polymer is thought of as being composed of the two words; "poly" meaning many and "mer" meaning unit. Thus, the word polymer means many units and is very descriptive of a polymer molecule. Several of these terms will be reexamined in this chapter but the intent of the remainder of this book is to principally consider aspects of the last three.

TYPES OF STRAIN ELLIPSES AND ELLIPSOIDS

Strain Ellipsoid

Let us refer again to the sphere inscribed within our cube that undergoes homogeneous strain. Giving this sphere axes l_1, l_2 and l_3, we find that these remain orthogonal through the homogeneous deformation and can be used to describe the long, medium, and short dimensions of the resulting ellipsoid.

In each case, the inscribed circle and ellipse represent cross sections through the strain ellipsoid before and after deformation. The types of strain shown are: (a) uniform extension; (b) uniform flattening; and (c) plane strain. Note that no strain occurs in the intermediate direction.)

Together, these are called the principal strain axes of the strain ellipsoid. Just as stress can be said to exist at every point within a body, so is there a corresponding strain ellipsoid for these points, once deformation has taken place. Thus, comparison between the "before" and "after" shape and axes of the sphere inscribed within our cube provides us with a measure of the amount and type of strain. It is standard practice for geologists to derive the principal axes of stress by superimposing them on the strain ellipsoid.

This is an optimistic simplification (as we have seen, the principal axes of stress and strain coincide only under conditions of homogeneity), but can be very useful. It has, for example, offered considerable insight into basic mechanisms and patterns of deformation-particularly faulting and fracturing-on many scales. This will become increasingly clear in the following two sections on folding and faulting. Because of their frequent use of structural cross sections, geologists have also found it advantageous to make use of a strain ellipse-essentially a cross section through the strain ellipsoid along the l_1 l_3 plane (i.e., the one that involves s_1 and s_3). The justification for this is, again, dependent on the assumption of homogeneous strain. Because of the regional nature of most tectonism, and the layered nature of lithologic sequences, many geologic examples of strain can be considered to approximate plane strain. In this type of deformation, the intermediate axis remains the diameter of the "original" sphere, while shortening and stretching occur along the other two axes.

Thus, two dimensions are sufficient to describe the strain at a particular point. If we are ready to accept the assumption of homogeneous strain, the strain ellipse becomes one of our principal indicators for the summation- of-local-strains method.

Some natural materials, such as ooids, spherulites, pebbles, certain fossils, and reduction spots in shales, can be used as qualitative ellipses or, in some cases, ellipsoids. However, because volume changes frequently occur during deformation (especially in carbonates) any quantitative determinations of strain based on such materials must be used with caution.

In principle, any object whose initial shape is known can act as a strain indicator. Such an indicator can be important to the subsurface explorationist, since it may be the only direct evidence available for how much strain has affected the fabric-and thus porosity and permeability-of a lithologic section. In most cases, the degree of tectonic influence on grain texture is fairly apparent from petrographic study. Strain indicators are primarily useful where this may not be clear and where special circumstances warrant mathematical determinations of strain. Specific techniques for measuring finite strain from oolites and spherulites are given by Ramsay (1967) and Ramsay and Huber (1985).

Nearly all deformation in nature is in-homogeneous. Not only do originally planar surfaces become complexly curved, but volume changes that involve both loss and addition of material frequently take place. Because of their pronounced heterogeneity in composition, thickness, and thus strength, rocks do not behave passively during deformation, but adjust in complex ways. Some units become strain-hardened and are able to withstand and transfer greater and greater amounts of stress as deformation progresses; other lithologies, in contrast, are fated to absorb stress by flowage, recrystallization, and the development of secondary fabrics such as cleavage.

Again, despite the dominance of inhomogeneity in nature, both local and regional deformational history can be reconstructed by assuming near-homogeneous strain domains. On a large scale, this often establishes the regional nature of stress and strain. Geologists often estimate a regional strain ellipse based on the orientation of major structural trends. This is called the mean strain ellipse and is often useful in explaining such trends in terms of plate interactions.

Types of Strain

Much of the terminology derived for understanding stress has also been applied to strain. Our hypothetical cube is said to have suffered homogeneous strain (also called uniform strain) when the strain is the same at all points within it. This means that originally straight lines remain straight after deformation.

This should also help make clear the basic concept of inhomogeneous strain, by far the most common in naturally deformed rock. This type of strain involves some amount of rotation in the position of particles, which means that originally straight lines become warped and detailed analysis becomes impracticable. It is, therefore, almost always useful to find some way in which natural deformation can be approximated as homogeneous. The most common approach is to consider geologic structures as the summation of many localized homogeneous strain fields. This method has proved especially helpful in the explanation of secondary rock fabrics, such as mineral alignment and fracturing.

Such fabrics often provide invaluable clues to both small- and large-scale structural patterns. Natural fracturing, of course, is of particular importance to petroleum geology, and knowledge of the stress-strain relationship associated with it can be very useful. As we will see, fracture patterns very often have a direct causal relation to major structures, such as faults and folds. The summation-of-local-strains method, therefore, will usually reveal this and permit the geologist to predict patterns in adjacent, undrilled areas. Several examples of this are given later on, when we look at fracturing in detail.

There are a number of basic ways in which deformation by homogeneous strain at constant volume occurs.

To understand how geologists treat natural deformation, however, it is also necessary that we look at the two basic types of shear strain. Both types help us explain a great many large- and small-scale features seen in rocks.

Pure shear is a form of strain in which no rotation of the strain axes takes place. It is often referred to as an irrotational deformation." It results from uniform extension in one direction and contraction perpendicular to it. Strain that approximates pure shear is seen in many folds.

In simple shear, all particles within a body are displaced in one direction. This is our cube pressed into a rhombohedron again; this time, however, we need to take note of the rotation in the strain ellipsoid. Simple shear can be visualized by imagining the result of placing our cube (with its inscribed sphere) between the two surfaces of an active fault. The shearing motion created by these two surfaces stretches and flattens the sphere into a strain ellipsoid whose long axis is progressively rotated until it is nearly parallel to the fault plane itself. Displacement within such a body takes place by slippage along closely spaced planes.

In actual materials, this can be accomplished in a number of ways, for example, by slippage between grains or crystals, or by actual flow at elevated temperatures and pressures.

TWO-DIMENSIONAL STRAIN AND STRESS ANALYSES

Stress Analysis

Stress analysis is an engineering (e.g., civil engineering and mechanical engineering) discipline that determines the stress in materials and structures subjected to static or dynamic forces or loads. A stress analysis is required for the study and design of structures, e.g., tunnels, dams, mechanical parts, and structural frames among others, under prescribed or expected loads. Stress analysis may be applied as a design step to structures that do not yet exist.

The aim of the analysis is usually to determine whether the element or collection of elements, usually referred to as a structure, can safely withstand the specified forces. This is achieved when the determined stress from the applied force(s) is less than the ultimate tensile strength, ultimate compressive strength or fatigue strength the material is known to be able to withstand, though ordinarily a factor of safety is applied in design.

Analysis may be performed through mathematical modelling or simulation, through experimental testing procedures, or a combination of techniques.

Engineering quantities are usually measured in megapascals (MPa) or gigapascals (GPa). In imperial units, stress is expressed in pounds-force per square inch (psi) or kilopounds-force per square inch (ksi).

Analysis methods: The analysis of stress within a body implies the determination at each point of the body of the magnitudes of the nine stress components. In other words, it is the determination of the internal distribution of stresses.

A key part of analysis involves determining the type of loads acting on a structure, including tension, compression, shear, torsion, bending, or combinations of such loads. When forces are applied, or expected to be applied, repeatedly, nearly all materials will rupture or fail at a lower stress than they would otherwise. The analysis to determine stresses under these cyclic loading conditions is termed fatigue analysis and is most often applied to aerodynamic structural systems.

Modelling: To determine the distribution of stress in a structure it is necessary to solve a boundary-value problem by specifying the boundary conditions, i.e. displacements and/or forces on the boundary. Constitutive equations, such as e.g. Hooke's Law for linear elastic materials, are used to describe the stress:strain relationship in these calculations. A boundary-value problem based on the theory of elasticity is applied to structures expected to deform elastically, i.e. infinitesimal strains, under design loads. When the loads applied to the structure induce plastic deformations, the theory of plasticity is implemented.

Approximate solutions for boundary-value problems can be obtained through the use of numerical methods such as the finite element method, the finite difference method, and the boundary element method, which are implemented in computer programs. Analytical or close-form solutions can be obtained for simple geometries, constitutive relations, and boundary conditions.

All real objects occupy a three-dimensional space. The stress analysis can be simplified in cases where the physical dimensions and the loading conditions allows the structure to be assumed as one-dimensional or two-dimensional. For a two-dimensional analysis a plane stress or a plane strain condition can be assumed.

Experimental testing: Stress analysis can be performed experimentally by applying forces to a test element or structure and then determining the resulting stress using sensors. In this case the process would more properly be known as *testing* (destructive or non-destructive). Experimental methods may be used in cases where mathematical approaches are cumbersome or inaccurate. Special equipment appropriate to the experimental method is used to apply the static or dynamic loading.

There are a number of experimental methods which may be used:

Tensile testing is a fundamental materials science test in which a sample is subjected to uniaxial tension until failure. The results from the test are commonly used to select a material for an application, for quality control, and to predict how a material will react under other types of forces. Properties that are directly measured via a tensile test are ultimate tensile strength, maximum elongation and reduction in area. From these measurements properties such as Young's modulus, Poisson's ratio, yield strength, and strain-hardening characteristics can be determined.

The photoelastic method relies on the physical phenomenon of birefringence. Unlike the analytical methods of stress determination, photoelasticity gives a fairly accurate picture of stress distribution even around abrupt discontinuities in a material. The method serves as an important tool for determining the critical stress points in a material and is often used for determining stress concentration factors in irregular geometries. Birefringence is exhibited by certain transparent materials. A ray of light passing through a birefringent material experiences two refractive indices. This double refraction is exhibited by many optical crystals. But photoelastic materials exhibit the property of birefringence only on the application of stress, and the magnitude of the refractive indices at each point in the material is directly related to the state of stress at that point. A model component is created made of photoelastic material with similar geometry to that of the structure on which stress analysis is to be performed. This ensures that the state of the stress in the model is similar to the state of the stress in the structure.

Dynamic mechanical analysis is a technique used to study and characterize viscoelastic materials, particularly polymers. Polymers composed of long molecular chains have unique viscoelastic properties, which combine the characteristics of elastic solids and Newtonian fluids. The viscoelastic property of polymer is studied by dynamic mechanical analysis where a sinusoidal force (stress) is applied to a material and the resulting displacement (strain) is measured. For a perfectly elastic solid, the resulting strain and the stress will be perfectly in phase. For a purely viscous fluid, there will be a 90 degree phase lag of strain with respect to stress. Viscoelastic polymers have the characteristics in between where some phase lag will occur during DMA tests.

Analyzers are made for both stress and strain control. In strain control, a probe is displaced and the resulting stress of the sample is measured. In stress control, a set force is applied and several other experimental conditions (temperature, frequency, or time) can be varied.

Factor of safety: The factor of safety is a design requirement for the structure based on the uncertainty in loads, material strength, and consequences of failure. In design of structures, calculated stresses are restricted to be less than an specified allowable stress, also known as working or designed stress, that is chosen as some fraction of the yield strength or of the ultimate strength of the material which the structure is made of. The ratio of the ultimate stress to the allowable stress is defined as the factor of safety.

Laboratory test are usually performed on material samples in order to determine the yield strength and the ultimate strength that the material can withstand before failure. Often a separate factor of safety is applied to the yield strength and to the ultimate strength. The factor of safety on yield strength is to prevent detrimental deformations and the factor of safety on ultimate strength is to prevent collapse.

Load transfer: The evaluation of loads and stresses within structures is directed to finding the load transfer path. Loads will be transferred by physical contact between the various component parts and within structures. The load transfer may be identified visually, or by simple logic for simple structures. For more complex structures, more complex methods such as theoretical solid mechanics or by numerical methods may be required. Numerical methods include direct stiffness method which is also referred to as the finite element method.

The object is to determine the critical stresses in each part, and compare them to the strength of the material.

For parts that have broken in service, a forensic engineering or failure analysis is performed to identify weakness, where broken parts are analysed for the cause or causes of failure. The method seeks to identify the weakest component in the load path. If this is the part which actually failed, then it may corroborate independent evidence of the failure. If not, then another explanation has to be sought, such as a defective part with a lower tensile strength than it should for example.

Uniaxial stress: If two of the dimensions of the object are very large or very small compared to the others, the object may be modelled as one-dimensional. In this case the stress tensor has only one component and is indistinguishable from a scalar. One-dimensional objects include a piece of wire loaded at the ends and a metal sheet loaded on the face and viewed up close and through the cross section.

When a structural element is subjected to tension or compression its length will tend to elongate or shorten, and its cross-sectional area changes by an amount that depends on the Poisson's ratio of the material. In engineering applications, structural members experience small deformations and the reduction in cross-sectional area is very small and can be neglected, i.e., the cross-sectional area is assumed constant during deformation. For this case, the stress is called *engineering stress* or *nominal stress*. In some other cases, e.g., elastomers and plastic materials, the change in cross-sectional area is significant, and the stress must be calculated assuming the current cross-sectional area instead of the initial cross-sectional area.

Plane Strain

If one dimension is very large compared to the others, the principal strain in the direction of the longest dimension is constrained and can be assumed as zero, yielding a plane strain condition. In this case, though all principal stresses are non-zero, the principal stress in the direction of the longest dimension can be disregarded for calculations. Thus, allowing a two dimensional analysis of stresses, e.g. a dam analyzed at a cross section loaded by the reservoir.

Stress and Strain Analysis and Measurement

The engineering design of structures using polymers requires a thorough knowledge of the basic principles of stress and strain analysis and measurement. Readers of this book should have a fundamental knowledge of stress and strain from a course in elementary solid mechanics and from an introductory course in materials. Therefore, we do not rigorously derive from first principles all the necessary concepts. However, in this chapter we provide a review of the fundamentals and lay out consistent notation used in the remainder of the text. It should be emphasized that the interpretations of stress and strain distributions in polymers and the properties derived from the standpoint of the traditional analysis given in this chapter are approximate and not applicable to viscoelastic polymers under all circumstances. By comparing the procedures discussed in later chapters with those of this chapter, it is therefore possible to contrast and evaluate the differences.

Some Important and Useful Definitions: In elementary mechanics of materials (Strength of Materials or the first undergraduate course in solid mechanics) as well as in an introductory graduate elasticity course five fundamental assumptions are normally made about the characteristics of the materials for which the analysis is valid. These assumptions require the material to be,

- Linear
- Homogeneous
- Isotropic
- Elastic
- Continuum.

Provided that a material has these characteristics, be it a metal or polymer, the elementary stress analysis of bars, beams, frames, pressure vessels, columns, etc. using these assumptions is quite accurate and useful. However, when these assumptions are violated serious errors can occur if the same analysis approaches are used. It is therefore incumbent upon engineers to thoroughly understand these fundamental definitions as well as how to determine if they are appropriate for a given situation. As a result, the reader is encouraged to gain a thorough understanding of the following terms:

Linearity: Two types of linearity are normally assumed: Material linearity (Hookean stress-strain behavior) or linear relation between stress and strain; Geometric linearity or small strains and deformation. Elastic: Deformations due to external loads are completely and instantaneously reversible upon load removal.

Continuum: Matter is continuously distributed for all size scales, i.e. there are no holes or voids.

Homogeneous: Material properties are the same at every point or material properties are invariant upon translation.

Inhomogeneous or Heterogeneous: Material properties are not the same at every point or material properties vary upon translation.

Amorphous: Chaotic or having structure without order. An example would be glass or most metals on a macroscopic scale.

Crystalline: Having order or a regular structural arrangement. An example would be naturally occurring crystals such as salt or many metals on the microscopic scale within grain boundaries.

Isotropic: Materials which have the same mechanical properties in all directions at an arbitrary point or materials whose properties are invariant upon rotation of axes at a point. Amorphous materials are isotropic.

Anisotropic: Materials which have mechanical properties which are not the same in different directions at a point or materials whose properties vary with rotation at a point. Crystalline materials are anisotropic.

Plastic: The word comes from the Latin word plasticus, and from the Greek words plastikos which in turn is derived from plastos (meaning molded) and from plassein (meaning to mold). Unfortunately, this term is often used as a generic name for a polymer probably because many of the early polymers (cellulose, polyesters, etc.) appear to yield and/or flow in a similar manner to metals and could be easily molded. However, not all polymers are moldable, exhibit plastic flow or a definitive yield point.

Viscoelasticity or Rheology: The study of materials whose mechanical properties have characteristics of both solid and fluid materials. Viscoelasticity is a term often used by those whose primary interest is solid mechanics while rheology is a term often used by those whose primary interest is fluid mechanics. The term also implies that mechanical properties are a function of time due to the intrinsic nature of a material and that the material possesses a memory (fading) of past events. The latter separates such materials from those with time dependent properties due primarily to changing environments or corrosion. All polymers (fluid or solid) have time or temperature domains in which they are viscoelastic.

Polymer: The word Polymer originates from the Greek word "polymeros" which means many-membered, (Clegg and Collyer 1993). Often the word polymer is thought of as being composed of the two words; "poly" meaning many and "mer" meaning unit. Thus, the word polymer means many units and is very descriptive of a polymer molecule. Several of these terms will be reexamined in this chapter but the intent of the remainder of this book is to principally consider aspects of the last three.

4

Igneous Rocks

Classification of igneous rocks is one of the most confusing aspects of geology. This is partly due to historical reasons, partly due to the nature of magmas, and partly due to the various criteria that could potentially be used to classify rocks.

- Early in the days of geology there were few rocks described and classified. In those days each new rock described by a geologist could have shown characteristics different than the rocks that had already been described, so there was a tendency to give the new and different rock a new name. Because such factors as cooling conditions, chemical composition of the original magma, and weathering effects, there is a potential to see an infinite variety of igneous rocks, and thus a classification scheme based solely on the description of the rock would eventually lead to a plethora of rock names. Still, because of the history of the science, many of these rock names are firmly entrenched in the literature, so the student must be aware of all of these names, or at least know where to look to find out what the various rocks names mean.
- Magmas, from which all igneous rocks are derived, are complex liquid solutions. Because they are solutions, their chemical composition can vary continuously within a range of compositions. Because of the continuous variation in chemical composition there is no easy way to set limits within a classification scheme.
- There are various criteria that could be used to classify igneous rocks. Among them are:
  1. Minerals Present in the Rock (the *mode*). The minerals present in a rock and their relative proportions in the rock depend largely on the chemical composition of the magma. This works well as a classification scheme if all of the minerals that could potentially crystallize from the magma have done so - usually the case for

slowly cooled plutonic igneous rocks. But, volcanic rocks usually have their crystallization interrupted by eruption and rapid cooling on the surface. In such rocks, there is often glass or the minerals are too small to be readily identified. Thus a system of classification based solely on the minerals present can only be used.

We can easily the inadequacy of a mineralogical classification based on minerals present if you look at the classification schemes for volcanic rocks given in introductory geology textbooks. For example, most such schemes show that a dacite is a rock that contains small amounts of quartz, somewhat larger amounts of sanidine or alkali feldspar, plagioclase, biotite, and hornblende, In all the years I have been looking at igneous rocks (since about the mid-cretaceous) I have yet to see a dacite that contains alkali feldspar. Does this mean that the intro geology textbooks lie? Not really, these are the minerals that should crystallize from a dacite magma, but don't because the crystallization history is interrupted by rapid cooling on the surface.

2. Texture of the Rock. Rock texture depends to a large extent on cooling history of the magma. Thus rocks with the same chemical composition and same minerals present could have widely different textures. In fact we generally use textural criteria to subdivide igneous rocks in to plutonic (usually medium to coarse grained) and volcanic (usually fine grained, glassy, or porphyritic.) varieties.
3. Color. Color of a rock depends on the minerals present and on their grain size. Generally, rocks that contain lots of feldspar and quartz are light colored, and rocks that contain lots of pyroxenes, olivines, and amphiboles (ferromagnesium minerals) are dark colored. But color can be misleading when applied to rocks of the same composition but different grain size. For example a granite consists of lots of quartz and feldspar and is generally light colored. But a rapidly cooled volcanic rock with the same composition as the granite could be entirely glassy and black colored (i.e. an obsidian). Still we can divide rocks in general into *felsic rocks* (those with lots of feldspar and quartz) and*mafic rocks* (those with lots of ferromagnesium minerals). But, this does not allow for a very detailed classification scheme.
4. Chemical Composition. Chemical composition of igneous rocks is the most distinguishing feature.
    - o The composition usually reflects the composition of the magma, and thus provides information on the source of the rock.

- o The chemical composition of the magma determines the minerals that will crystallize and their proportions.
- o A set of hypothetical minerals that could crystallize from a magma with the same chemical composition as the rock (called the *Norm*), can facilitate comparison between rocks.
- o Still, because chemical composition can vary continuously, there are few natural breaks to facilitate divisions between different rocks.
- o Chemical composition cannot be easily determined in the field, making classification based on chemistry impractical.

Because of the limitations of the various criteria that can used to classify igneous rocks, geologists use an approach based on the information obtainable at various stages of examining the rocks.

1. In the field, a simple field based classification must be used. This is usually based on mineralogical content and texture. For plutonic rocks, the IUGS system of classification can be used.
2. Once the rocks are brought back to the laboratory and thin sections can be made, these are examined, mineralogical content can be more precisely determined, and refinements in the mineralogical and textural classification can be made.

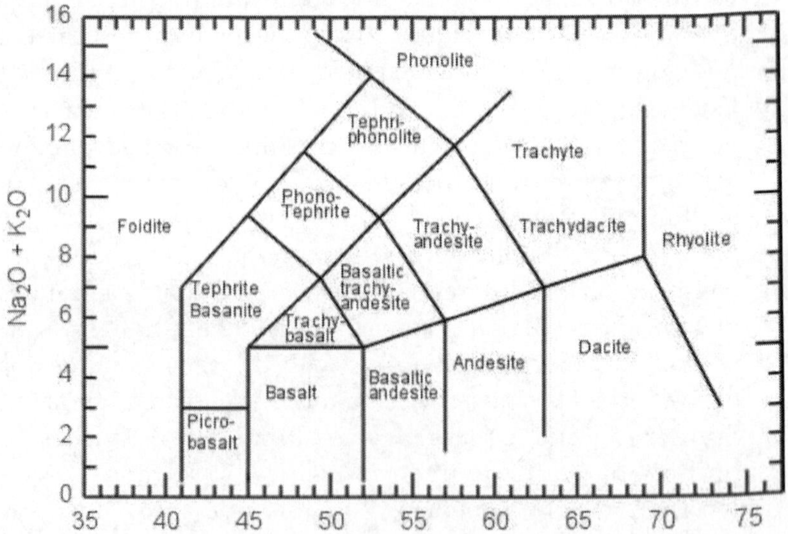

3. Chemical analyses can be obtained, and a chemical classification, such as the LeBas et al., IUGS chemical classification of volcanic rocks

Note that at each stage of the process, the classification may change, but it is important to keep in mind that each stage has limitations, and that classification at each stage is for the purposes of describing the rock, not only for the individual investigator, but anyone else. Thus, the classification scheme should be employed in a consistent manner so that later investigators can understand what you are talking about at each stage of the process.

General Chemical Classifications

SiO_2 (Silica) Content
> 66 wt. % - Acid
52-66 wt% - Intermediate
45-52 wt% - Basic
< 45 wt % - Ultrabasic

This terminology is based on the onetime idea that rocks with a high % SiO_2 were precipitated from waters with a high concentration of hyrdosilicic acid H_4SiO_4. Although we now know this is not true, the acid/base terminology is well entrenched in the literature.

Silica Saturation

If a magma is oversaturated with respect to Silica then a silica mineral, such as quartz, cristobalite, tridymite, or coesite, should precipitate from the magma, and be present in the rock. On the other hand, if a magma is undersaturated with respect to silica, then a silica mineral should not precipitate from the magma, and thus should not be present in the rock. The silica saturation concept can thus be used to divide rocks in silica undersaturated, silica saturated, and silica oversaturated rocks. The first and last of these terms are most easily seen.

- Silica Undersaturated Rocks - In these rocks we should find minerals that, in general, do not occur with quartz. Such minerals are:

Nepheline- $NaAlSiO_4$
Leucite - $KAlSi_2O_6$
Forsteritic Olivine - Mg_2SiO_4
Sodalite - $3NaAlSiO_4 \cdot NaCl$
Nosean - $6NaAlSiO_4 \cdot Na_2SO_4$
Haüyne - $6NaAlSiO_4 \cdot (Na_2,Ca)SO_4$
Perovskite - $CaTiO_3$
Melanite - $Ca_2Fe^{+3}Si_3O_{12}$
Melilite - $(Ca,Na)_2(Mg,Fe^{+2},Al,Si)_3O_7$

Thus, if we find any of these minerals in a rock, with an exception that we'll see in a moment, then we can expect the rock to be silica undersaturated.

If we calculate a CIPW Norm the normative minerals that occur in silica undersaturated rocks are nepheline and/or leucite.

- Silica Oversaturated Rocks. These rocks can be identified as possibly any rock that does notcontain one of the minerals in the above list.

If we calculate a CIPW Norm, silica oversaturated rocks will contain normative quartz.

- Silica Saturated Rocks. These are rocks that contain just enough silica that quartz does not appear, and just enough silica that one of the silica undersaturated minerals does not appear. In the CIPW norm, these rocks contain olivine, or hypersthene + olivine, but no quartz, no nepheline, and no leucite.

To get an idea about what silica saturation means, let's look at a simple silicate system - the system Mg_2SiO_4 - SiO_2

Note how compositions between Fo and En will end their crystallization with only Fo olivine and enstatite.

These are SiO_2-undersaturated. compositions. All compositions between En and SiO_2 will end their crystallization with quartz and enstatite. These are SiO_2 - oversaturated compositions.

Note also that this can cause some confusion in volcanic rocks that do not complete their crystallization due to rapid cooling on the surface. Let's imagine first a composition in the silica-undersaturated field. Cooling to anywhere on the liquidus will result in the crystallization of Fo-rich olivine. If this liquid containing olivine is erupted and the rest of the liquid quenches to a glass, then this will produce a rock with phenocrysts of olivine in a glassy groundmass.

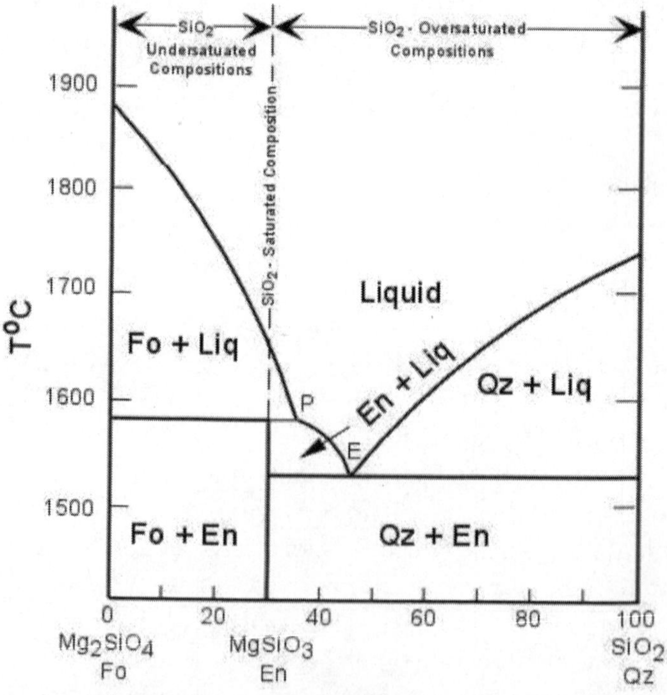

Applying the criteria above for identifying silica undersaturated rocks would tell us that this is a silica-undersaturated rock, which we know to be correct. Next, let's look at a silica oversaturated composition, such as one just to the left of the point labeled 'P'.

If this liquid is cooled to the liquidus and olivine is allowed to crystallize, and is then quenched on the surface, it will contain phenocrysts of Fo-rich olivine in a glassy groundmass.

Applying the criteria above would suggest that this rock is also silica undersaturated, but we know it is not. This illustrates one of the difficulties of applying any criteria of classification to volcanic rocks where incomplete crystallization/reaction has not allowed all minerals to form.

Alumina (Al_2O_3) Saturation

After silica, alumina is the second most abundant oxide constituent in igneous rocks. Feldspars are, in general, the most abundant minerals that occur in igneous rocks. Thus, the concept of alumina saturation is based on whether or not there is an excess or lack of Al to make up the feldspars. Note that Al_2O_3 occurs in feldspars in a ratio of 1 Al to 1 Na, 1K, or 1 Ca:

$KAlSi_3O_8$ — $1/2K_2O : 1/2Al_2O_3$

$NaAlSi_3O_8$ — $1/2Na_2O : 1/2Al_2O_3$

$CaAl_2Si_2O_8$ — $1CaO : 1Al_2O_3$

Three possible conditions exist.

1. If there is an excess of Alumina over that required to form feldspars, we say that the rock is*peraluminous*. This condition is expressed chemically on a molecular basis as:

   $Al_2O_3 > (CaO + Na_2O + K_2O)$

   In peraluminous. rocks we expect to find an Al_2O_3-rich mineral present as a modal mineral - such as muscovite [$KAl_3Si_3O_{10}(OH)_2$], corundum [Al_2O_3], topaz [$Al_2SiO_4(OH,F)_2$], or an Al_2SiO_5- mineral like kyanite, andalusite, or sillimanite.

   Peraluminous rocks will have corundum [Al_2O_3] in the CIPW norm and no diopside in the norm.

2. *Metaluminous* rocks are those for which the molecular percentages are as follows:

   $Al_2O_3 < (CaO + Na_2O + K_2O)$ and $Al_2O_3 > (Na_2O + K_2O)$

   These are the more common types of igneous rocks. They are characterized by lack of an Al_2O_3-rich mineral and lack of sodic pyroxenes and amphiboles in the mode.

3. *Peralkaline* rocks are those that are oversaturated with alkalies ($Na_2O + K_2O$), and thus undersaturated with respect to Al_2O_3. On a molecular basis, these rocks show:

   $Al_2O_3 < (Na_2O + K_2O)$

Peralkaline rocks are distinguished by the presence of Na-rich minerals like aegerine [$NaFe^{+3}Si_2O_6$], riebeckite [$Na_2Fe_3^{+2}Fe_2^{+3}Si_8O_{22}(OH)_2$], arfvedsonite [$Na_3Fe_4^{+2}(Al,Fe^{+3})Si_8O_{22}(OH)_2$], or aenigmatite [$Na_2Fe_5^{+2}TiO_2Si_6O_{18}$] in the mode.

In the CIPW norm, acmite [$NaFe^{+3}Si_2O_6$] and/or sodium metasilicate Na_2SiO_3 will occur as normative minerals.

IGNEOUS ROCK CLASSIFICATION

Igneous rocks form by crystallization during cooling, which can occur quickly to produce fine-grained rocks or more slowly to produce coarse-grained rocks. Because of these two different rates, above ground and fast versus below ground and slow, igneous rocks are found to occur in textural pairs. That is, for any given mineralogical composition it is possible to have two different textures. Igneous rocks formed from minerals at the base of Bowen's reaction series (mica, Na-plageoclase, K-feldspar and quartz) are known as felsic. The coarse-grained felsic rock granite is the principle lithologic component of the continental crust. Granite's fine-grained compositional equivalent, rhyolite, is formed in volcanic eruptions within and along the margins of continents. Conversely, igneous rocks formed from minerals near the top of Bowen's reaction series (olivine, Ca-plagioclase and pyroxene) are known as mafic.

The fine-grained mafic rockbasalt is the most abundant volcanic rock on Earth and forms the floor of most of the world's oceans. Basalt's coarse-grained compositional equivalent, gabbro, makes up the deeper portions of the oceanic crust. Between these two end-members there exist rocks rich in amphibole and mixed Ca-Na-plageoclase feldspars. These intermediate composition rocks are known as andesite (fine-grained) and diorite (coarse-grained). When an igneous rock contains minerals from only the top of Bowen's reaction series (essentially only olivine) the composition is said to be ultramafic. Because of the great density of ultramafic magmas fine-grained ultramafic rocks are very rare, the magma is simply too dense to reach the surface and cool quickly. Therefore, ultramafic rocks are only represented by the coarse-grained rock peridotite.

Defining the textural characteristics and the mineral composition of an igneous rock allows for categorization of that rock in terms of the igneous rock classification system. This system does more than provides a specific name for a group of similar rocks; rather, it provides the intellectual framework for a deeper understanding of all igneous rocks. By combining the Bowen's reaction series concept with the classification system, one is able to describe similarities and differences between and among the various groups of igneous rocks.

GEOMETRY OF IGNEOUS INTRUSIONS

Intrusive igneous rocks are, by volume, the most important type of rock found on Earth. Because of their great abundance, these intrusive rocks have been the subject of detailed analysis for several hundred years. While early theories suggested that these coarse-grained rocks might have formed by precipitation from seawater, we now know that in fact they form by crystallization from magmas deep underground. This mode of formation, cooling under the Earth's surface, has lead to a great deal of study of the geometries of intrusive igneous rock bodies.

The principle criterion upon which the geometries of igneous intrusions are classified is the nature of their relationships with surrounding rock. While some melts form in place and remain fixed as they cool, most magmas move during and after formation. Because of the lower density of hot molten magma relative to surrounding solid rock, magmas tend to rise through the Earth. As a melt moves upward, it interacts with the previously formed rocks of that region – known as country rock.

IGNEOUS ROCKS OF THE CONVERGENT MARGINS

The convergent plate margins are the most intense areas of active magmatism above sea level at the present time. Most of world's violent volcanic activity occurs along these zones. In addition, much magmatism also has resulted (and probably is resulting at present) in significant additions to the crust in the form of plutonic igneous rocks. Here, we look at this magmatism in terms of the volcanic rocks that appear to be related to subduction.

PETROGRAPHY

Probably the most distinguishing feature of subduction-related volcanic rocks is their usually porphyritic nature, usually showing glomeroporphyritic clusters of phenocrysts. Basalts commonly contain phenocrysts of olivine, augite, and plagioclase.

Andesites and dacites commonly have phenocrysts of plagioclase, augite, and hypersthene, and some contain hornblende. The most characteristic feature of the andesites and dacites is the predominance of fairly calcic plagioclase phenocrysts that show complex oscillatory zoning. Such zoning has been ascribed to various factors, including:

- Kinetic factors during crystal growth. As one zone of the crystal is precipitated the liquid immediately surrounding the crystal becomes depleted in the components necessary for further growth of the same composition. So, a new composition is precipitated until diffusion has had time to renourish the surrounding liquid in the components necessary for the equilibrium composition to form.
- Cycling through a chemically zoned magma chamber during convection. As crystals grow, they are carried in convection cells to warmer and cooler parts of the magma chamber. Some zones are partially dissolved and new compositions are precipitated that are more in equilibrium with the chemical compositions, pressures, and temperatures present in the part of the magma chamber into which the crystal is transported.
- Magma mixing. As magmas mix the chemical compositions of liquids and temperatures change during the mixing process. This could result in dissolution of some zones, and precipitation of zones with varying chemical composition.

Rhyolites occur as both obsidians and as porphyritic lavas and pyroclastics. Phenocrysts present in rhyolites include plagioclase, sanidine, quartz, orthopyroxene, hornblende, and biotite.

In addition to these features, petrographic evidence for magma mixing is sometimes present in the rocks, including disequilibrium mineral assemblages, reversed zoning etc. Xenoliths of crustal rocks are also sometimes found, particularly in continental margin arcs, suggesting that assimilation or partial assimilation of the crust could be an important process in this environment.

Major Elements

Before discussing the major element chemistry of subduction related volcanic rocks we first need to clarify some terminology concerning rock suites. In the early 1900s a petrologist by the name of Peacock examined suites of rocks throughout the world. On a plot of CaO and total alkalies versus SiO_2, Peacock noted that the two curves intersected at different values of SiO_2 for different suites. He used the value of SiO_2 where the two curves intersect (now known as the *Peacock Index* or *Alkali-Lime Index*) to divide rock suites into the following:

Peacock Index	*Name of Suite*
<51	Alkalic
51-56	Alkali-Calcic
56-61	Calc-Alkalic
>61	Calcic

Although Peacock's classification of rock suites is rarely used today, some of the terminology has survived in slightly different forms. For example the general term "alkaline suite" is used to describe rock suites in which the basic rocks have relatively high values of total alkalies, like the alkali basalt - hawaiite - mugearite - trachyte suite or the basanite - nephelinite suites discussed previously. Most subduction related volcanic and plutonic rocks fall into the calc-alkalic suite of Peacock, and thus the term calc-alkaline is often given to the suite of rocks found associated with subduction. But, it is notable that all four suites of rocks defined by Peacock are found in subduction-related areas.

The modern definition of the calc-alkaline suite is based on the AFM. The subduction-related volcanic rocks show a trend along which the ratio of MgO to total iron (MgO/FeO*) remains nearly constant. This trend is often referred to as the *Calc-Alkaline trend*. Note that the calc-alkaline trend is distinct from the Fe-enrichment trends shown by the alkaline and tholeiitic suites discussed previously. Also, calc-alkaline basalts, like tholeiitic basalts are subalkaline, but they differ from tholeiitic basalts in their higher concentrations of Al_2O_3, with values of 17 to 20 wt %. Thus, these calc-alkaline basalts are often referred to as *high alumina basalts.*

Recall that the Fe-enrichment trend exhibited by the tholeiitic and alkaline rock series can be explained by crystal fractionation involving removal of early crystallizing Mg-rich olivines and pyroxenes from the parental basaltic magmas. The calc-alkaline trend, however, would require early removal of mineral assemblages with a higher Fe/Mg ratio, or some other process. Over the years several explanations for the calc-alkaline trend have been discussed. Among these are:

- Crystal fractionation by early removal of an Fe-rich mineral assemblage. Because the basaltic compositions are similar to tholeiitic basalts, they would crystallize the same Mg-rich olivines and pyroxenes as tholeiitic basalts. So, this process would require early crystallization of additional Fe-rich phases to raise the Fe/Mg ratio of the early crystallizing assemblage. Likely candidates for the Fe-rich phase or phases would be magnetite or an Fe-rich amphibole. Experiments conducted in the 60s through 80s failed to show that magnetite or Fe-rich amphibole would be early crystallizing phases in basalts or andesites under geologically reasonable conditions. So, at least initially, this mechanism appeared to be unacceptable.

- Assimilation of crustal material by basaltic magmas. Since rhyolites and granites are chemically similar, and since the continental crust contains a higher proportion of granitic rocks, it could be possible that the calc-alkaline trend is due to assimilation of crustal granites by basaltic magmas. The difficulty of such a process operating on a large scale because of the energy requirements involvéd. A larger hindrance to this mechanism, however, is that the calc-alkaline suite occurs both in island arcs, where there is little or no continental crust, as well as in continental margin arcs where there is such crust.
- Magma Mixing. The calc-alkaline trend could be explained by mixing of basaltic magmas with rhyolitic magmas to produce the intermediate andesites and dacites. This involves the problem of first, how are the rhyolites generated, and second that such rhyolitic and basaltic magmas would have to be present beneath all arcs. While mixing does seem to play a role, it is unlikely that it always occurs and is always able to generate the large volumes of magma required to build a mostly andesitic stratovolcano.
- Andesites as primary magmas. In the early years, when it was not recognized that basalts do occur in the arcs or at least that andesites were the predominant type of magma erupted, it was suggested that andesites were primary magmas. Since it was known that the mantle would not likely be able to produce silica oversaturated andesitic magma by partial melting, except at very low pressure, it was suggested that the subducted oceanic crust partially melted to produce andesitic magmas. This seemed like a good hypothesis in light of the new theory of Plate Tectonics that was coming out at the time. But, as we will see later, there are serious obstacles to this theory in the trace element composition of the magmas. Nevertheless, this early theory became popular and was put into introductory physical geology textbooks, many of which still advocate that andesitic magmas are generated by partial melting of the subducted oceanic crust.

In recent years more light has been shed on the possible origin of the calc-alkaline suite. Perhaps the best evidence comes from experimental petrology and recent advances in experimental techniques. Experimental petrology has long suffered from the possibility that the experimental charge could possibly react with the container in which it was placed.

Thus, the choice of the container or capsule, as it is called, is very important. Perhaps the best in this regard is gold. Gold remains relative inert at high temperatures, and thus does not appear to react with silicate liquids in any major way. But, the melting temperature of Gold is about 1060°C which means that experiments must be conducted at temperatures that are relatively low compared with those of basaltic liquids.

Platinum (Pt) has a much higher temperature and is inert and does not react with liquids that have no Fe. But Pt absorbs Fe from liquids if it is present. Thus, if one is attempting to determine whether or not an Fe-rich phase crystallizes from a liquid at high temperature, Fe-loss to the Pt capsule could become important (i.e. it might suppress the crystallization of an Fe-rich phase because there is less Fe in the liquid than would be present under natural conditions).

Although this limitation was recognized and attempts were made in early experiments to minimize Fe loss to the Pt, the experiments still remained suspect. In the 1980's, however, techniques were developed to saturate the Pt capsules with Fe prior to experimentation. This led to important new experiments addressing the problem of the calc-alkaline suite.

First, however, consider experiments conducted at low pressure on tholeiitic basalt magmas. These experiments show that at low pressure Plagioclase and Olivine crystallize first, with proportion of plagioclase crystallizing being higher than that of olivine. On the projected phase diagram, removal of Olivine and Plagioclase drives the liquid composition away from the Olivine corner until it intersects the Ol + Plag + Cpx + Liq. cotectic. Further crystallization of these phases will then drive the liquid composition along the cotectic to eventually crystallize pigeonite (low Ca-pyroxene) as shown by the light colored path on the diagram.

Analyses of the liquids produced in these experiments showed that, as expected the liquids would follow a trend of Fe-enrichment and thus the calc-alkaline trend could not be produced by fractional crystallization at low pressures.

Next, experiments were conducted at a pressure of 2 kb with enough H_2O in the capsules to assure that the liquid would be H_2O saturated at this pressure (i.e. a free vapor phase would coexist with the liquid). These experiments were conducted because it was known that H_2O would lower the temperature of appearance of the silicate minerals, but would lower the temperature of appearance of oxide minerals, like magnetite to a lesser extent, and could stabilize a hydrous phases like hornblende at a higher temperature.

The experiments show that

- The position of the Ol + Plag + Cpx + Liq. cotectic shifts toward the Olivine corner of the projected phase diagram.
- The proportion of Olivine relative to plagioclase becomes much higher than in the dry low pressure experiments.
- Magnetite becomes an early crystallizing phase and hornblende also crystallizes early if the liquids have a high enough concentration of Na_2O.
- Most importantly, analyses of the liquids produced in the experiments plot along the calc-alkaline trend in the AFM.

Furthermore, if subduction related arc rocks are plotted on the projection there are seen to lie in a field surrounding the 2 kb H_2O saturated cotectic. This indicates that the calc-alkaline suite could be produced by fractional crystallization under moderate pressure water saturated conditions. This would suggest that the main difference between tholeiitic rocks and calc-alkaline rocks might be the presence (in calc-alkaline basalts) or absence (in tholeiitic basalts) of H_2O in the parental magmas and/or the source rocks that melt. We know that it is possible to introduce water into the subduction related environment by dehydration of the subducting lithosphere, whereas it is more difficult to envision a mechanism to add water to the source where tholeiitic magmas are generated.

5

Landslide Hazards and Disasters

BHACHAU LANDSLIDE

The land slipped during the 26th January 2001 earthquake event in Bhachau. Note people are still camped beneath. Monsoon rains could possibly wash the soil downslope.

Human Excavation of slope and its toe, Loading of slope/toe, draw down in reservoir, mining, deforestation, irrigation, vibration/blast, Water leakage from services.

Earthquake shaking has triggered landslides in many different topographic and geologic settings. Rock falls, soil slides and rockslides from steep slopes involving relatively thin or shallow dis-aggregated soils or rock, or both have been the most abundant types of landslides triggered by historical earthquakes. Volcanic eruption Deposition of loose volcanic ash on hillsides commonly is followed by accelerated erosion and frequent mud or debris flows triggered by intense rainfall.

Elements at Risk

The most common elements at risk are the settlements built on the steep slopes, built at the toe and those built at the mouth of the streams emerging from the mountain valley. All those buildings constructed without appropriate foundation for a given soil and in sloppy areas are also at risk. Roads, communication line and buried utilities are vulnerable.

Indian Landslides

Landslide constitute a major natural hazard in our country, which accounts for considerable loss of life and damage to communication routes, human settlements, agricultural fields and forest lands. The Indian subcontinent, with diverse physiographic, seismotectonic and climatological conditions is subjected to varying degree of landslide hazards; the Himalayas including

Northeastern mountains ranges being the worst affected, followed by a section of Western Ghats and the Vindhyas. Removal of vegetation and toe erosion have also triggered slides Torrential monsoon on the vegetation cover removed slopes was the main causative factors in the Peninsular India namely in Western Ghat and Nilgiris. Human intervention by way of slope modification has added to this effect.

Hazard Zones

The Landslide Hazard Zonation Map of India is produced in the next page. The Landslide Map needs to be popularized among the architects, engineers and development planners and also to the public so that it is used as a tool for regulating construction or development activities and means of managing or mitigating landslide disasters.

Typical Effects

Physical Damage: Landslides destroy anything that comes in their path. They block or bury roads, lines of communication, settlements, river flow, agricultural land, etc. It also includes loss to agricultural production and land area. In addition physical effects such as flooding may also occur. The Malpa village as on 17 August 1998, a few hours before the rock avalanche wiped out the village around 0300hours on the 18 August, 1998. The two pictures were taken from the same location upstream of Kali River. They provide a direct comparison of scenarios before and after the event.

Landslide in Kerala

In the Kerala part of the Western Ghats several types of mass movements /landslides have been recorded. The most prevalent, recurring and disastrous type of mass movements noted in Kerala are the "debris flows". The swift and sudden down slope movement of highly water saturated overburden containing a varied assemblage of debris material ranging in size from soil particles to huge boulders destroying and carrying with it every thing that is lying in its path. Note the boulder movement along the path.

Casualties: They cause maximum fatalities depending on the place and time of occurrence. Catastrophic landsides have killed many thousands of persons, such as the debris slide on the slopes of Huascaran in Peru triggered by an earthquake in 1970, which killed over 18,000 people.

Main Mitigation Strategies

Hazard mapping will locate areas prone to slope failures. This will permit to identify avoidance of areas for building settlements. These maps will serve as a tool for mitigation planning.

Land use practices such as:

Areas covered by degraded natural vegetation in upper slopes are to be afforested with suitable species. Existing patches of natural vegetation (forest and natural grass land) in good condition, should be preserved. Any developmental activity initiated in the area should be taken up only after a detailed study of the region and slope protection should be carried out if necessary.

In construction of roads, irrigation canals etc. proper care is to be taken to avoid blockage of natural drainage

Total avoidance of settlement in the risk zone should be made mandatory.

Relocate settlements and infrastructure that fall in the possible path of the landslide No construction of buildings in areas beyond a certain degree of slope.

Retaining Walls can be built to stop land from slipping (these walls are commonly seen along roads in hill stations). It's constructed to prevent smaller sized and secondary landslides that often occur along the toe portion of the larger landslides.

Surface Drainage Control Works. The surface drainage control works are implemented to control the movement of landslides accompanied by infiltration of rain water and spring flows.

Engineered structures with strong foundations can withstand or take the ground movement forces. Underground installations (pipes, cables, etc.) should be made flexible to move in order to withstand forces caused by the landslide Increasing vegetation cover is the cheapest and most effective way of arresting landslides. This helps to bind the top layer of the soil with layers below, while preventing excessive run-off and soil erosion. Insurance will assist individuals whose homes are likely to be damaged by landslides or by any other natural hazards. For new constructions it should include standards for selection of the site as well as construction technique.

Community based Mitigation

The most damaging landslides are often related to human intervention such as construction of roads, housing and other infrastructure in vulnerable slopes and regions. Other community based activities that can mitigate landslides are education and awareness generation among the communities, establishing community based monitoring, timely warning and evacuation system.

Communities can play a vital role in identifying the areas where there is land instability. Compacting ground locally, slope stabilization (procedures such as terracing and tree planting may reduce damages to some extent), and avoiding construction of houses in hazardous locations are something that the community has to agree and adhere to avoid damage from the possible landslides. This would also reduce the burden of shifting of settlements from hazardous slopes and rebuild in safe site as it is less practical to do in large scale.

Debris Flows or Mudslies

Fast-moving flows of mud and rock, called debris flows or mudslides, are among the most numerous and dangerous types of landslides in the world. They are particularly dangerous to life and property because of their high speeds and the sheer destructive force of their flow.

Hazardous Areas

Debris flows start on steep slopes-slopes steep enough to make walking difficult. Once started, however, debris flows can travel even over gently sloping ground. The most hazardous areas are canyon bottoms, stream channels, areas near the outlets of canyons, and slopes excavated for buildings and roads.

A: Canyon bottoms, stream channels, and areas near the outlets of canyons or channels are particularly hazardous. Multiple debris flows that start high in canyons commonly funnel into channels. There, they merge, gain volume, and travel long distances from their sources.

B: Debris flows commonly begin in swales (depressions) on steep slopes, making areas downslope from swales particularly hazardous.

C: Roadcuts and other altered or excavated areas of slopes are particularly susceptible to debris flows. Debris flows and other landslides onto roadways are common during rainstorms, and often occur during milder rainfall conditions than those needed for debris flows on natural slopes.

D: Areas where surface runoff is channeled, such as along roadways and below culverts, are common sites of debris flows and other landslides.

Onset Type

They strike suddenly although it takes time to build up. They can be tracked on the development but accurate landfall is predictable barely few hours. The onset is extensive and often very destructive.

HAZARDOUS PROCESSES

What evidence exists for an increase in the frequency and/or magnitude of potentially damaging processes such as floods, debris flows, and landslides? The catastrophic debris torrent events of 1994 occurred in Phojal Nalla, a medium-sized tributary valley and stream on the right bank of the Beas River between the towns of Kullu and Manali. An examination of the morphology and sedimentology of its lower reaches and confluence with the Beas River revealed evidence of repeated high-magnitude torrents and floods in the form of a boulderystreambed (bedload) and coarse-caliber stream-bank levees, features that are diagnostic of very high magnitude flows. These indications are confirmed by historical evidence of hazard events. A catastrophic debris torrent occurred in Phojal Nalla in June 1894.

As many as 200 people and several thousand head of livestock were swept away. Damage to property crops, trails, and other infrastructure was extensive, and the morphology of the Beas River channel was altered downstream from the confluence. Upstream investigations at the time and described in press reports indicated no deforestation or other significant land-cover change, but they did note a large slope failure and damming of the stream course that contributed to the flood. Content analysis of old documents and newspapers did not indicate an event of similar magnitude at this location until 1994. The latter event was attributed to a "cloudburst" during the monsoon season.

In 1894 the victims were largely migratory herders and traders and their livestock, whereas in 1994 the victims were people who had built residences and small businesses along or in the stream course. Morphological and sedimentological evidence of extreme flows is present in many of the first-order channels throughout the area. This is supported by copious written commentary on frequent destructive floods and debris flows from the latter part of the nineteenth century and the early twentieth century. Further support comes from a review of the Civil and Military Gazette, in which death, damage, and disruptions from these processes were reported on a regular basis, especially during the monsoon season.

Many of the currently occupied fans and terraces that form the present agricultural land base comprise debris flow and flood deposits, providing evidence of very high magnitude paraglacial erosion and deposition processes. A high-magnitude debris flow that blocked the Solang River and caused flooding occurred as recently as August 2001. The same debris fan is the subject of a legend which depicts an enormous "deluge" in the distant past—certainly plausible given the fan morphology and sedimentology—that destroyed all the villages and all their inhabitants but one. The survivor, a young girl, ran to the top of a small but still prominent hill that protrudes above the debris. Of course, she then became the mother of the reoccupation of the destroyed land by succeeding generations (Banon 1995). Again, there is no evidence of recent deforestation or degradation in the catchment of this stream course and fan.

Morphology and sedimentology indicate that high-magnitude flows have occurred regularly on the Beas River, and the historical record indicates that floods have caused damage to property deaths, and injuries in the human and livestock populations, as well as severe bank erosion. There is no physical or historical evidence that floods have increased in frequency or magnitude. Indeed, Thomas Hofer found that the annual range between low and high flows was decreasing, suggesting a more stable annual hydrograph—the opposite of what would be expected under conditions of upstream deforestation (1993). The active channel, incised into the paraglacial fan and terrace deposits, is wide and composed of bouldery, coarse-caliber material across which the stream shifts, indicative of high energy associated with steep gradient and highly variable discharge.

Over time, some areas stabilise and, with continued stream incision, become isolated from annual floods and may be occupied. During high-magnitude floods, as happened in August and September 1995, such areas have been eroded and/or inundated, damaging property, roads, and structures. This appears to have been the pattern throughout the historical period. Likewise, landslide processes are not new to the area, though an event in September 1995 at Luggar Bhatti near Kullu town, which killed sixty-five people, led many to believe that they were becoming more frequent. Although less in evidence than flood and debris-flow landforms and deposits, large-magnitude landslide features are present on and at the base of many valley-side slopes.

Again, folklore provides some circumstantial evidence of catastrophic landslides in the past that is corroborated by present-day landforms. Near Bandrole, on the right bank of the Beas River upstream from Kullu town, is evidence of an ancient landslide. In the early travel literature on the area mention is made of an old woman who sought shelter in a village near this location; shelter was denied, and she was sent on to stay in the tents of traders camped nearby.

That night a landslide destroyed the village and all of its inhabitants. The encampment that gave refuge to the old woman was spared, and the woman, subsequently referred to as "the spirit of the mountain," was nowhere to be found! Other notable landslide features are found on the left bank of the Beas, by Manali and at the village of Solang, upstream from Manali. Historical photographs of the former indicate its presence in the nineteenth century, and photographic and field comparisons today indicate relative stability. Solang is situated on a progressive landslide feature on the left bank of the Solang River. Tension cracks and multiple slump scars indicate continuing, though slow, failure, possibly related to undercutting by the river. Finally, descriptions of the collateral effects throughout the Kullu District of the 1905 Kangra earthquake include commentary about numerous landslides and boulder deposits, from Larji in the south to Naggar, north of Kullu town (Punjab

District Gazetteer 1918). Similar collateral landslides are described in Garhwal in association with the 1991 earthquake.

Nothing exists in the written or physical record to indicate that, in general, these potentially hazardous processes are increasing in frequency, magnitude, or spatial distribution. Furthermore, in the upper Beas River watershed no substantial evidence exists for widespread reduction in forest cover or other radical destructive land-surface change that could explain an increased frequency or magnitude of potentially hazardous processes during the historical period. Saczuk indicates nothing from the climatic record that could lead to an increase in process occurrence (2001). The 1995 Luggar Bhatti disaster, however, does point to a new phenomenon that leads to increased frequency of relatively low-magnitude land instabilities. This refers to anthropogenic disturbance during and after the construction of major roads and highways in the area—a phenomenon that is well documented in other parts of the Himalaya.

The construction of roads across steep slopes, which is unavoidable in the Kullu District, as in other mountain areas, destabilises the local slope through removal of the toe slope in sediments and steepening of rock slopes. Landslides and slips and rockfalls pose a continual hazard and maintenance problem on roads, especially during spring thaws and monsoons. Where roads are switchbacked upslope to gain altitude, as at Rohtang Pass, localised instability is extended over a larger area, such that the whole mountainside becomes unstable.

The geotechnical antecedents of the Luggar Bhatti disaster included the undercutting of an already unstable, erosionally steepened slope in glaciofluvial deposits during road construction, coupled with large infusions of moisture during monsoon rainfall. The fact that road construction was under way ensured the presence of a large number of construction workers and an elevated level of vulnerability. This anthropogenically stimulated landslide and socially constructed disaster perhaps provides a clue as to the recent and emerging perceptions of elevated risk from natural processes in this part of the Himalaya.

VULNERABILITY TO NATURAL HAZARDS

Vulnerability is defined by the degree of exposure of people, property, and infrastructure to dangerous processes and events through a juxtaposition in time and location. The remarkable growth of permanent and transient populations, structures, roads, and bridges, power and communications links, vehicles, investment-intensive land uses, and services in the Kullu District, especially since 1990, by itself increases exposure or vulnerability; other things being equal. Although there is no evidence of an increase in frequency and magnitude of most potentially damaging processes over time, likewise there is no evidence of a reduction in frequency and magnitude within the historical period. The public rhetoric of the 1990s recognises increased risk even though the reasons for it may not be clearly understood.

Characteristic of most high mountain areas, the Kullu District has limited land that is suitable for intensive agriculture and horticulture, for building structures, roads, and other communications links, and for expansion of settlements. The best and safest land was occupied decades, if not centuries, ago and is reflected in the distribution of the traditional villages. With the exception of more recent settlements such as Manali and the expanded Kullu town, the traditional communities have protected their land, including the village-use and Protected Forest areas, further limiting the growth of built-up areas to marginal lands along stream courses, at roadsides, and at the base of steep slopes.

Because of population increase, the number of structures in the core villages has increased, and this expansion has increased exposure to dangerous processes. Since 1970 the building of residences away from the core village areas in a scattered pattern, usually upslope, has begun to take hold, creating a mix of agricultural land and single residences. This pattern began at a small scale following the land redistribution in the 1970s. Saczuk has measured a 35 percent increase in built-up area over the 1972-1999 period (2001). Much of this occurred on the margins—and the margins are more dangerous.

State-level land-use regulation and zoning in flood-prone areas along the Beas River have been developed, but implementation and enforcement are inconsistent. And, as with all after-the-fact nonstructural adjustments to natural hazards, little can be done with existing and well-established structures and land uses in the hazard-prone areas in the short term. The August-September 1995 flood on the Beas River, the most damaging in a decade, redefined some of the flood-prone landscape by destroying or severely damaging structures and facilities within and outside the legislated flood-hazard zone upstream from Kullu town.

Many of the destroyed and damaged structures have not been rebuilt, but major artificial levees have been constructed to protect the remaining structures and facilities, including an important military and snow avalanche research centre, and any future structures in the hazard-prone area. In the absence of careful monitoring and enforcement in flood-hazard zones, the same or different land uses will creep back into these and other exposed areas. Perhaps the factors that have the greatest restraining influence in rebuilding in hazard-prone areas such as this are lack of capital and absence of damage insurance.

Most encouraging is the Himachal Cabinet prohibition of January 2001 on further building adjacent to the Beas River upstream from Manali, despite the fact that the stimulus for this was to control growth and the need for services, rather than to reduce vulnerability. To be effective, however, serious enforcement of the regulation will have to occur. Some properties owned by national and multinational corporations and various levels of government were able to survive the 1995 flood with little or no damage and have persisted.

Their presence in the flood-prone areas, in contradiction to zoning regulations, maintains a high level of vulnerability and has been challenged through public-interest litigation in the High Court of Himachal Pradesh. Even this, when successful, has not led to removal of the offending structures and land uses. Away from the Beas River and its flood-prone areas, there has been little recognition of hazard zones through detailed hydrogeophysical analysis and mapping, backed by zoning legislation and building restrictions. Exposure to landslides, debris flows, and floods in the tributary valleys and on the fans and terraces adjacent to the Beas River remains.

In Manali, larger and newer hotels have been built on marginal land immediately adjacent to the Manalsu Nalla. Some of these hotels were damaged by the 1995 flood. Other ready examples are found upstream on the left bank of the Beas River, where a range of settlements, migratory-worker camps, small businesses, small hotels and guest houses, amusement parks, and major government facilities such as the Snow and Avalanche Study Establishment, are exposed to debris-flow hazard from upslope gullies and to flood hazard from the Beas. Recognising that many of the land uses in hazard-prone areas cannot be relocated and that land for future building is extremely limited, the state and local governments have undertaken major physical structural adjustments to protect land from a variety of hazardous processes.

These retaining and restraining devices or walls are usually made up of gabion baskets and, in some cases, earth fill. They are used as dikes to retain flood flows in predefined channels, as in-channel walls to prevent bank erosion and some spillover, and as retaining walls at the base of steep and mobile slopes that intersect roads. Most of the structural adjustments in the Kullu District have been made since July 1993, when early monsoon rain and snow at higher elevations produced significant flooding on the Beas River and its tributaries. The most persistent hazards in the Kullu District are those associated with roads, especially NH21.

During snowmelt in the spring and during the summer monsoon season (July-September), the roads are especially vulnerable to small slides, washouts, rockfalls, debris flows, and, in some cases, major slope failures. Although slope failures produce the major catastrophes, the greatest impacts are by way of frequent road closures and traffic disruptions. Given the strategic and commercial importance and escalating use of NH21, these closures and disruptions, some of which may last as long as a week, are not only annoying but also have substantial economic impacts and implications. The growing dependence on road transport of the economy of the area and the region thus escalates the level of vulnerability.

The roads are extremely difficult to maintain and keep open, giving rise to another set of adjustments that is characteristic of India and South Asia. This form of adjustment is based on the presence of a very large force of road-maintenance workers who are supported by state, municipal, and army personnel with machinery for road clearance.

When not cleaning up the aftermath of a landslide, rockfall, or washout on the road, the workforce is employed in ongoing road maintenance throughout the area. This is a migratory workforce, which joins that generated by the tourist and horticultural industries and which includes families and extended families, usually from Uttar Pradesh, Bihar, or Nepal. Its presence in the area creates its own social problems of housing, water supply, and waste disposal quite apart from the vulnerability inherent in the work itself, which was so tragically demonstrated in the Luggar Bhatti disaster.

The lack of adequate and safe housing for these and other migratory workers and their families leads to the establishment of densely populated camps in hazardous areas along the roadways themselves and adjacent to dangerous stream channels and nallas.

This so-called floating population in the Kullu-Manali area numbers more than 10,000 and is present during the most hazardous time of the year, the summer monsoon season. The combination of circumstances produces a high level of vulnerability for a population that has little recourse to aid and resources in the event of a disaster.

The old trans-Himalayan trade that brought traders and merchants to encampments along the Beas River and exposed them to events such as the 1894 Phojal Nalla disaster is now dead. In its place is a globally driven economy that brings a transient, much larger population of tourists and migrant workers to the area to face the same dangers. By definition, vulnerability has increased.

Proclivities, not Certainties

This research supports findings in a growing number of studies that the Himalayan environmental degradation theory does not apply. Like most areas in the Himalaya, the Kullu District has its own physical and cultural-historical characteristics that have shaped its landscape and human ecology. This study provides an example from the Indian Himalaya, where the British administration left an indelible imprint on land-use and forest-management practices and institutions, an imprint that was relatively conservative with respect to deforestation, as has been demonstrated elsewhere. Forest-coverage and forest-area boundaries in the upper Beas River watershed were approximately the same in 2001 as in the late nineteenth century; with a slight reduction in forested area since 1972 and some infilling of cleared land by forest over a longer period. Large-scale and complete deforestation is nowhere evident.

The pattern of working forests is relatively unchanged, though some nontimber and timber forest products are less accessible. These findings reflect three factors: the traditional presence of village-use areas, which included forests, in the Kullu District; the entering into law of some rights of use in Protected Forests under colonial forest settlements; and persistence of the village-based society and economy in the Kullu District in the face of strong regional and global forces of change. Landforms and sediments of varying age and information in various historical documents suggest that flood, debris flow, and landslide processes have operated in similar locations, with similar frequency and magnitude, over long periods of time. Large, paraglacial debris flow and alluvial fans and terraces provide evidence of very high magnitude processes in the early postglacial period. With the exception of one fan, there is no evidence of occurrences of similar magnitude during the period of human habitation.

Although the data are scanty and more archival work is required, at this point there is no evidence for an increase in processes, with the exception of failures of steepened slopes at road cuts. The notion of an environment degraded by deforestation and overgrasing and giving rise to increased occurrence of dangerous processes is not supported. However, further research on the frequency and magnitude of geomorphic and hydrological processes and their relationships to meteorological, geological, ecological, and anthropogenic conditions is warranted. Risk from natural hazards is present, as documentation of particularly catastrophic floods, debris flows, and landslides demonstrates. Although no link to deforestation is demonstrated and process occurrence shows no measurable increase except at road cuts, vulnerability of people and property has increased. This is explained by growth of permanent and transient populations, expansion of built-up areas, and construction and upgrading of roads and other communications infrastructure, often in areas that are exposed to hazardous processes, here as elsewhere. Mountain environments have proclivities for extreme natural processes, but disaster events in the Kullu District increasingly are constructed from social, economic, and political ingredients of local, regional, and global origin.

TSUNAMI HAZARDS AND DISASTERS

TSUNAMI - is a Japanese word meaning "harbour wave". These waves, which often affect distant shores, originate from undersea or coastal seismic activity, landslides, and volcanic eruptions. Whatever the cause, sea water is displaced with a violent motion and swells up, ultimately surging over land with great destructive power.

26th December, 2004, A massive earthquake of Magnitude 9.0 hit Indonessia generating Tsunami waves in South-east Asia & eastern coast of India. Height of tsunami waves ranged from 3 - 10 m affecting a total coastal length of 2260 km in the States of Andhra Pradesh, Tamil Nadu, Kerala & UTs of Pondicherry, Andaman & Nicobar Islands.

Tsunami waves travelled upto a depth of 3 km from the coast killing more than 10,000 people & affected more than lakh of houses leaving behind a huge trail of destruction.

Onset Type and Causes

If the earthquake or under water land movement is near the coast then tsunami may strike suddenly and if the earth movement is far in the sea then it may take few minutes to hours before striking the coast. The onset is extensive and often very destructive. The general causes of Tsunamis are geological movements. It is produced in three major ways. The most common of these is fault movement on the sea floor, accompanied by an earthquake. To say that an earthquake causes a tsunami is not completely correct. Rather, both earthquakes and tsunamis result from fault movements.

Probably the second most common cause of tsunamis is a landslide either occurring underwater or originating above the sea and then plunging into the water. The third major cause of tsunamis is volcanic activity. The flank of a volcano, located near the shore or underwater, may be uplifted or depressed similar to the action of a fault. Or, the volcano may actually explode. In 1883, the violent explosion of the famous volcano, Krakatoa in Indonesia, produced tsunamis measuring 40 meters which crashed upon Java and Sumatra. over 36,000 people lost their lives as a result of tsunami waves from Krakatoa. The giant tsunamis that are capable of crossing oceans are nearly always created by movement of the sea floor associated with earthquakes which occur beneath the sea floor or near the ocean.

Warning

Tsunami is not a single giant wave. It consists of ten or more waves which is termed as a "tsunami wave train".

Since scientists cannot predict when earthquakes will occur, they cannot predict exactly when a tsunami will be generated. Studies of past historical tsunamis indicate where tsunamis are most likely to be generated, their potential heights, and flooding limits at specific coastal locations. With use of satellite technology it is possible to provide nearly immediate warnings of potentially tsunamigenic earthquakes. Warning time depends upon the distance of the epicenter from the coast line.

The warning includes predicted times at selected coastal communities where the tsunami cold travel in a few hours. In case of tsunamigenic earthquakes or any other geological activity people near to the coastal areas may get very little time to evacuate on receiving of warning.

Elements at Risk

All structures located within 200 m of the low lying coastal area are most vulnerable to the direct impact of the tsunami waves as well as the impact of debris & boulders brought by it. Settlements in adjacent areas will be vulnerable to floods & scour.

Structures constructed of wood, mud, thatch, sheets and structures without proper anchorage to foundations are liable to be damaged by tsunami waves & flooding.

Other elements at risk are infrastructure facilities like ports & harbours, telephone and electricity poles, cables. Ships & fishing boats/nets near the coast also add to the destruction caused by tsunami waves.

Typical Effects

Physical Damage : Local tsunami events or those less than 30 minutes from the source cause the majority of damage. The force of can raze everything in its path. It is the flooding effect of a tsunami, however, that most greatly effects human settlements by water damage to homes and businesses, roads, bridges and other infrastructure. Ships, port facilities, boats/trawlers, fishing nets also get damaged.

Environmental Damage : There is evidence of ever increasing impact upon the environment on account of the effects of tsunamis. The range varies from generation of tonnes of debris on account of structural collapse of weaker buildings, release of toxic chemicals into the environment on account of chemical leak/spillage/process failure/utility breakages/ collateral hazards and negative impact on the already fragile ecosystems.

Casualties and Public Health: Deaths occur principally from drowning as water inundates homes or neighborhoods. Many people may be washed out to sea or crushed by the giant waves. There may be some injuries from battering by debris and wounds may become contaminated.

Water supply: sewage pipes may be damaged causing major sewage disposal problems. Drinking water shortage arises due to breakage of water mains and contamination. Open wells and ground water may become unfit for drinking due to contamination of salt water and debris.

Standing Crops and food supplies: flooding by tsunami causes damage to the standing crops and also to the food supplies in the storage facilities. The land may be rendered infertile due to salt water incursion from the sea.

Specific Preparedness Measures

Hazard mapping - a hazard map should be prepared with designated areas expected to be damaged by flooding by tsunami waves. Historical data could be of help ion showing areas inundated in the past. Keeping in mind the vulnerable areas, evacuation routes should be constructed and mapped. The plan should be followed by evacuation drill.

Early warning systems - a well networked system in place can warn the communities of the coastal areas when the threat is perceived. Tsunami warning should be disseminated to local, state, national as well as the international community so as to be prepared as they are capable of crossing across continents. The information can be broadcasted to the local emergency officers and the general public. On receiving of the warning the action should be to evacuate the place as decided in the evacuation plan. Community Preparedness - communities in the coastal areas are faced by the wrath of cyclones, storm surge and tsunami waves. It is important that the community is better prepared to take suitable actions on receiving of the threat and follow emergency evacuation plans and procedures. a community which choose to ignore warning may get severely effected if they are not prepared to take immediate measures.

Main Mitigation Strategies

Site Planning and Land Management: Within the broader framework of a comprehensive plan, site planning determines the location, configuration, and density of development on particular sites and is, therefore, an important tool in reducing tsunami risk. The designation and zoning of tsunami hazard areas for such open-space uses as agriculture, parks and recreation, or natural hazard areas is recommended as the first land use planning strategy. This strategy is designed to keep development at a minimum in hazard areas.

In areas where it is not feasible to restrict land to open-space uses, other land use planning measures can be used. These include strategically controlling the type of development and uses allowed in hazard areas, and avoiding high-value and high-occupancy uses to the greatest degree possible.

The capital improvement planning and budgeting process can be used to reinforce land use planning policies.

Engineering Structures: As most of the structures along the coast area comprises of fisherman community, which are constructed of light weight materials without any engineering inputs. Therefore there is an urgent need to educate the community about the good construction practices that they should adopt such as:

Site Selection: Avoid building or living in buildings within several hundred feet of the coastline as these areas are more likely to experience damage from tsunamis. Construct the structure on a higher ground level with respect to mean sea level.

Elevate Coastal Homes: Most tsunami waves are less than 3 meters in height. Elevating house will help reduce damage to property from most tsunamis. Structural columns resist the impact while other walls are expendable. It is important to also take note that adequate measures are also brought into the design to cater for earthquake forces. Construction of water breakers to reduce the velocity of waves. Use of water & corrosion resistant materials for construction.

Construction of community halls at higher locations, which can act as shelters at the time of a disaster.

Flood Management: Flooding will result from a tsunami. Tsunami waves will flood the coastal areas.

Flood mitigation measures could be incorporated.

Building barriers or buffers such as special breakwaters or seawalls can be an effective risk reduction measure against gushing waters in case of Tsunami/Storm surge during cyclones.

Onset Type and Warning

Though they occur gradually, however sudden failure (sliding) can occur without warning. They may take place in combination with earthquakes, floods and volcanoes. There are no clearly established warnings in place indicating occurrence of landslide and hence difficult to predict the actual occurrence. Areas of high risk can be determined by use of information on geology, hydrology, vegetation cover, past occurrence and consequences in the region.

Causes Of Landslides

Geological Weak Material: Weathered materials, jointed or fissured materials, contrast in permeability and contrast in stiffness (stiff, dense material over plastic materials).

Erosion: Wave erosion of slope toe, glacial erosion of slope toe, subterranean erosion (Deposition loading slope or its crest, Vegetation removal.

Intense Rainfall: Storms that produce intense rainfall for periods as short as several hours or have a more moderate intensity lasting several days have triggered abundant landslides.

EARTHQUAKE CODES

Scientists have been able to identify the range of earthquakes of different magnitudes and different intensities, which could be, expected in different parts of the country, and also to some extent the probable frequency of occurrence of the highest range of earthquake in a specific region. This knowledge forms the basis of earthquake codes, which divide the country in different zones where earthquakes of a specific magnitude and a specific intensity could be expected.

The Codes also give guidelines for design of buildings with different materials, of different shapes and forms, and with different structural systems. Considering the unpredictable nature of earthquake, in its occurrence, its magnitude, its intensity and its duration, the Codes are very guarded in undertaking complete responsibility of the safety of a structure, even if it is designed following all its provisions rigorously. This is a stand taken by the codes all over the world, not only just in India. The philosophy of design adopted by all these codes is, that if a building is designed properly and constructed properly on the basis of the code, it should not suffer any damage under a mild earthquake, should suffer damage of only non-structural elements and finishes etc, which can easily be repaired under an earthquake of medium to high intensity, and should suffer damage of structural elements, without collapse, under very severe earthquake.

None of the codes states that if a building is designed following provisions of the code, and built properly, nothing will happen to it under any earthquake that can possibly occur in that region. Because those who frame these codes realize that if they had to make this statement, the cost of the buildings would be prohibitive. The buildings would have to be designed for earthquakes of very high intensity and magnitude, the probability of occurrence of which may be only once in fifty years, or once in hundred years.

It is interesting to trace the history of the process of making these earthquake codes. Early codes were based directly on the practical lessons learned from earthquakes, relating primarily to types of construction. In some cases they placed limitations on the height of buildings. It was recognized that timber structures performed well (even the relatively tall Japanese pagodas), whereas plain masonry buildings performed poorly. This process started in Italy way back in 1783 when a severe earthquake in "Calabrian" prompted the engineers to think of Earthquake (EQ) resistant buildings. On the basis of observations, the engineers stipulated that:

* All buildings which had failed & survived be built with timber frame, in-filled with stone embedded in mortar
* The maximum height of buildings be two storeys

However, these stipulations were not rigorously followed as the years, decades and a century went by, during which period there was extensive seismic activity in the region but of small to medium intensity and magnitude. Later in 1908 a severe earthquake occurred in "Messina-Reggio", during which 160,000 persons lost their lives in a relatively small area. Most of the collapsed buildings were in masonry and had not followed the stipulations framed in 1784.

A fter this earthquake, a commission of nine practicing engineers and five distinguished university professors were assigned the task of identifying methods to design buildings which were cheap, could be built easily an could also resist earthquakes. The commission gave two proposals:

1. Isolate the building from the ground and place it on a compacted layer of sand or on spherical rollers.
2. Construct the building with timber frame and in-built rubble masonry but design it to withstand horizontal force equal to 1/12 of its dead weight. This force was later changed to a horizontal design load of 1/12 of the weight above for the ground floor and 1/8 of the weight above for 2nd & 3rd floor.

Proposal (2) was generally adopted. These were intuitive recommendations based on observations and these concepts of designing a building to withstand a stipulated horizontal force or isolating it from the ground are still valid.

Japan also has a long history of earthquakes, which had intuitively led them to construct very light buildings in timber. Some of the wooden pagodas constructed before 15th century are amongst the tallest wooden structures in the world and have withstood many earthquake without any reported damage. Here again, a commission was set up which observed that buildings built in wood and steel had fared much better than those in concrete. But those in masonry had fared very badly.

It is interesting to note that three buildings designed by one Dr. Tachu Naito, Professor of Architecture at Waseda University in Tokyo, withstood the 1923 earthquake remarkably well. These three buildings were the Japan industrial Bank, 100ft high in steel frame, Jitsugyo Building in reinforced concrete frame, and the Kalenki Theatre in a combination of concrete and steel. All these buildings had been designed to withstand a lateral force equal to 1/15 of their dead weight. By 1880 a Seismological Programme had been set up and some empirical criteria for design of earthquake had been developed.

In 1923 a very severe earthquake took as many as 1,40,000 lives in Japan. By then, a number of building in steel and reinforced concrete had also been built, most of which withstood the earthquake fairly well. On account of success of these buildings to resist earthquake forces, Dr. Naito was considered an authority on the subject. He had laid down four very simple principles to be followed for earthquake resistant buildings:

1. A building should be as rigid as possible with rigid joints and generous bracings or shear walls. This will ensure short building period and prevent resonance with ground motion.
2. Use a closed plan layout, rather than an open U, L, T or H shapes.
3. Rigid walls or bracings should be placed symmetrically in plan.
4. Lateral force allocation to different frames of a building is done based on their rigidities.

The famous Imperial Hotel designed by Frank Lloyd Wright survived this 1923 earthquake in Japan without too much damage. It was a fairly rigid two storied building supported on short 8' long piles at 2'x2' grid. In the USA, a very severe earthquake occurred in 1906 in San Francisco. But it did not result in setting up any commission to make recommendations. Instead, the regular building code made a provision to design all buildings for a horizontal load of 30 pounds per square foot to cater for wind and earthquake forces.

It was only at 1925 "Santa-Barbara" earthquake that work to frame a seismic design code was undertaken which resulted in a "Uniform Building Code" published in 1927. The provisions in this code were, however, casual, to cater for a specified horizontal force, and were not mandatory. A subsequent "Long-Beach" earthquake in 1933, made the authorities serious. Laws were enacted and codes were framed. Serious provisions for designing the buildings for specified horizontal loads were made, which were modified over the years as more and more knowledge of earthquake engineering was acquired, and which became more and more stringent.

1933 - Important buildings like schools, hospitals, places of assembly for 10% (Dead + some Live load) all other buildings 2% to 5% (Dead + some Live load)

1943 - Horizontal load each floor = C x dead load above C=0.6/(N+4.5), N= no of stories above. Thus for a one storey building (N=0) C=0.133 and a 13 storeys building (N=12) C=0.0364

1947 - Horizontal load varies form 3.7 to 8% of design vertical load depending on number of storeys and soil conditions

1948 - Base shear V=CW

W=dead load +0.25 live load

C=0.015/T

T= fundamental period = 0.05H

H= height in ft.

D= plan dimension in the direction of earthquake in ft.

1956 - V same as above, but C=0.02/T

A number of subsequent revisions took place. The latest uniform building code is of 1997 edition. The name of this code has been changed to "International Building Code of USA" in 2000, the latest edition of which is of 2003.

The first Indian code for design of EQ resistant buildings was framed in 1962, which was subsequently revised in 1966, 1970, 1975, 1984 and finally in 2002. With every revision, revisions were modified on the basis of the latest available knowledge. However with every such revision provisions of the code became more stringent.

Originally, the country had been divided in 7 zones. Starting with EQ of very mild intensity in Zone-0, the intensity kept on increasing in Zones- I, II, III, IV and V with the heaviest in Zone-VI.

Subsequently in 1984 these seven zones were reduced to five. Zone-0 was merged into Zone-I and Zone-VI in Zone-V. Recently in 2002 the zones have further been reduced to four. Zone-I has been merged into Zone-II. The latest seismic zoning map of India showing four different zones (zone II, III, IV and V). The magnitude of seismic force experienced by a building varies with its configuration, construction materials, height, and number of floors, structural system, and type of foundation and soil characteristics.

It is directly related to the intensity of earthquake. Actual forces that a structure is subjected to during an earthquake may be far greater than those specified in this Code. However, ductility, arising from inelastic material behavior and detailing and over strength arising from the additional reserve strength in a structure, over and above the design strength, are relied upon to withstand these additional forces.

The strength requirement of a building to withstand earthquake-generated forces can be assessed based on a "Static Approach" or a "Dynamic Approach".

In the static approach, it is assumed that earthquake vibrations subject a building to horizontal forces along its height. The magnitude and distribution of such horizontal forces is related to the construction materials, height, and number of floors, foundation system, soil characteristics and the Earthquake zone in which the building is located. The total design lateral load is termed as the design seismic base shear.

In the Dynamic Approach, vibration analysis of the building is carried out to establish the base shear and its distribution over the height of the building. The Code specifies that all buildings can be designed with the static approach, except for buildings higher than 40m in Zones IV & V and higher than 90m in zones- II and III which need to be designed with the dynamic approach. But if a building is irregular it has to be designed according to dynamic approach, if it is higher than 12m in Zones-IV and V and higher than 40m in zones- II & III.

The Code has very stringent provisions to cater for torsion effects of earthquake forces. It also has equally stringent provisions to analyze and design irregular buildings.

EARTHQUAKES HAZARDS AND DISASTERS

The same process, results in earthquakes in the Andaman & Nicobar Islands. Sometimes earthquakes of different magnitudes occur within the Indian Plate, in the peninsula and in adjoining parts of the Arabian Sea or the Bay of Bengal. These arise due to localized systems of forces in the crust sometimes associated with ancient geological structures such as in the Rann of Kachchh. All earthquakes in peninsula India falls within this category.

Hazard Zones

As per the latest seismic zoning map of India the country is divided into four Seismic Zones. Zone V marked in red shows the area of Very High Risk Zone, Zone IV marked in orange shows the area of High Risk Zone. Zone III marked in yellow shows the region of Moderate Risk Zone and Zone II marked in blue shows the region of Low risk Zone. Zone V is the most vulnerable to earthquakes, where historically some of the country's most powerful shock has occurred.

Geographically this zone includes the Andaman & Nicobar Islands, all of North-Eastern India, parts of north-western Bihar, eastern sections of Uttaranchal, the Kangra Valley in Himachal Pradesh, near the Srinagar area in Jammu & Kashmir and the Rann of Kutchh in Gujarat. Earthquakes with magnitudes in excess of 7.0 have occurred in these areas, and have had intensities higher than IX.

Much of India lies in Zone III, where a maximum intensity of VII can be expected. New Delhi lies in Zone IV whereas Mumbai and Chennai lie in Zone III. All states and UTs across the country have experienced earthquakes.

Measuring the Size of an Earthquake - MSK 64 Intensity Scale

Intensity is a qualitative measure of the actual shaking at a location during an earthquake, and is notated in a roman capital numeral. The MSK (Medvedev, Sponheuer and Karnik) scale is more convenient for application in field and is widely used in India. The zoning criterion of the map is based on likely intensity.

The scale range from I (least perceptive) to XII (most severe). The intensity scales are bas ed on three features of shaking - perception by people, performance o f buildings, and changes to natural surroundings.

The seismic zoning map broadly classifies India into zones where one can expect earthquake shaking of the more or less the same maximum intensity.

Frightening

(i) Felt by most indoors and outdoors. Many people in buildings are frightened and run outdoors. A few persons loose their balance. Domestic animals rum out of their stalls. In many instances, dishes and glassware may break, and books fall down, pictures move, and unstable objects overturn. Heavy furniture may possibly move and small steeple bells may ring.

(ii) Damage of Grade 1*** is sustained in single** buildings of Type B* and in many** of Type A*. Damage in some buildings of Type A is of Grade 2***.

(iii) Cracks up to widths of 1cm possible in wet ground; in mountains occasional landslips: change in flow of springs and in level of well water are observed.

Damage of Buildings

(i) Most people are frightened and run outdoors. Many find it difficult to stand. The vibration is noticed by persons driving motor cars. Large bells ring.

(ii) In many buildings of Type C* damage of Grade 1 is caused; in many buildings of Type B damage is of Grade 2. Most** buildings of Type A suffer damage of Grade 3***, few of Grade 4***. In single instances, landslides of roadway on steep slopes: crack in roads; seams of pipelines damaged; cracks in stone walls.

(iii) Waves are formed on water, and is made turbid by mud stirred up. Water levels in wells change, and the flow of springs changes. Some times dry springs have their flow resorted and existing springs stop flowing. In isolated instances parts of sand and gravelly banks slip off.

Destruction of Buildings

(i) Fright and panic; also persons driving motor cars are disturbed, Here and there branches of trees break off. Even heavy furniture moves and partly overturns. Hanging lamps are damaged in part.

(ii) Most buildings of Type C suffer damage of Grade 2, and few of Grade 3, Most buildings of Type B suffer damage of Grade 3. Most buildings of Type A suffer damage of Grade 4. Occasional breaking of pipe seams. Memorials and monuments move and twist. Tombstones overturn. Stone walls collapse.

(iii) Small landslips in hollows and on banked roads on steep slopes; cracks in ground up to widths of several centimeters. Water in lakes becomes turbid. New reservoirs come into existence. Dry wells refill and existing

General Damage of Buildings

(i) General panic; considerable damage to furniture. Animals run to and fro in confusion, and cry.

(ii) Many buildings of Type C suffer damage of Grade 3, and a few of Grade 4. Many buildings of Type B show a damage of Grade 4 and a few of Grade 5. Many buildings of Type A suffer damage of Grade 5. Monuments and columns fall. Considerable damage to reservoirs; underground pipes partly broken. In individual cases, railway lines are bent and roadway damaged.

(iii) On flat land overflow of water, sand and mud is often observed. Ground cracks to widths of up to 10 cm, on slopes and river banks more than 10 cm. Further more, a large number of slight cracks in ground; falls of rock, many land slides and earth flows; large waves in water. Dry wells renew their flow and existing wells dry up.

Typical Effects

Physical Damage - damage or loss of buildings and service structures. Fires, floods due to dam failures, landslides could occur.

Casualties - often high, near to the epicenter and in places where the population density is high (say, multistoried buildings) and structures are not resistant to earthquake forces.

Public health - multiple fracture injuries, moderately and severely injured is the most widespread problem, breakdown in sanitary conditions and large number of casualties could lead to epidemics.

Water supply - severe problems due to failure of the water supply distribution network and storage reservoirs. Fire hydrants supply lines if vulnerable could hamper fire service operations.

Transport network - severely affected due to failure of roads and bridges, railway tracks, failure of airport runways and related infrastructure.

Electricity and Communication - all links affected. Transmission towers, transponders, transformers collapse.

Main Mitigation Strategies

Engineered structures (designed and built) to withstand ground shaking. Architectural and engineering inputs put together to improve building design and construction practice. Analyze soil type before construction and do not build structures on soft soil. To accommodate on weak soils adopt safety measures in design. Note: Buildings built on soft soils are more likely to get damaged even if the earthquake is not particularly strong in magnitude.

Similar problem persists in the alluvial plains and conditions across the river banks. Heavy damages are concentrated when ground is soft.

- Follow Indian Standard Codes for construction of buildings.
- Enforcement of the Byelaws including Land use control and restriction on density and heights of buildings
- Strengthening of important lifeline buildings which need to be functional after a disaster. Upgrade level of safety of hospital, fire service buildings etc.
- Public awareness, sensitization and training programmes for Architects, Builders, Contractors, Designers, Engineers, Financiers, Government functionaries, House owners, Masons etc.
- Reduce possible damages from secondary effects such as like fire, floods, landslides etc. e.g. identify potential landslide sites and restrict construction in those areas.

Community Based Mitigation

Community preparedness along with public education is vital for mitigating the earthquake impact. Earthquake drills and Public awareness programme.

Community based Earthquake Risk Management Project should be developed and sustainable programmes launched. Retrofitting of schools and important buildings, purchase of emergency response equipment and facilities, establishing proper insurance can be the programmes under Earthquake Risk Management Project. A large number of local masons and engineers will be trained in disaster resistant construction techniques. A large number of masons, engineers and architects can get trained in this process.

6

Environmental Conservation and Ecology

INTRODUCTION

Environmental conservation is an integral part of the socio economic development. The growing population, high degree of mechanisation and steep rise in energy use has led to activities that directly or indirectly affect the sustainability of the environment. The various kinds of pollution/pollutants may be broadly categorized as (a) Air pollution, (b) Water pollution, (c) Solid waste and (d) Industrial and hazardous waste. The sustainable use of bio-diversity is fundamental to ecologically sustainable development. India is one of the 12 mega diversity countries of the world. The survival and well being of any nation depends on sustainable social and economic progress that satisfies the needs and aspiration of the present without compromising the interest of future generations. Environmental problems are largely the by-products of affluence marked by resource wasteful life style. Recycling of solid and liquid wastes, bio-composting, tree planting, etc. are important for environmental conservation.

THE CONCEPT OF ECOSYSTEM

An ecosystem consists of the biological community that occurs in some locale, and the physical and chemical factors that make up its non-living or abiotic environment. There are many examples of ecosystems — a pond, a forest, an estuary, grassland. The boundaries are not fixed in any objective way, although sometimes they seem obvious, as with the shoreline of a small pond. Usually the boundaries of an ecosystem are chosen for practical reasons having to do with the goals of the particular study.

The study of ecosystems mainly consists of the study of certain processes that link the living, or biotic, components to the non-living, or abiotic, components. Energy transformations and biogeochemical cycling are the main processes that comprise the field of ecosystem ecology. As we learned earlier, ecology generally is defined as the interactions of organisms with one another and with the environment in which they occur. We can study ecology at the level of the individual, the population, the community, and the ecosystem.

Studies of individuals are concerned mostly about physiology, reproduction, development or behavior, and studies of populations usually focus on the habitat and resource needs of individual species, their group behaviors, population growth, and what limits their abundance or causes extinction. Studies of communities examine how populations of many species interact with one another, such as predators and their prey, or competitors that share common needs or resources.

In ecosystem ecology we put all of this together and, insofar as we can, we try to understand how the system operates as a whole. This means that, rather than worrying mainly about particular species, we try to focus on major functional aspects of the system. These functional aspects include such things as the amount of energy that is produced by photosynthesis, how energy or materials flow along the many steps in a food chain, or what controls the rate of decomposition of materials or the rate at which nutrients are recycled in the system.

The term ecosystem was coined in 1935 by the Oxford ecologist Arthur Tansley to encompass the interactions among biotic and abiotic components of the environment at a given site. The living and non-living components of an ecosystem are known as biotic and abiotic components, respectively.

Ecosystem was defined in its presently accepted form by Eugene Odom as, "an unit that includes all the organisms, i.e., the community in a given area interacting with the physical environment so that a flow of energy leads to clearly defined tropic structure, biotic diversity and material cycles, i.e., exchange of materials between living and non-living, within the system".

Smith (1966) has summarized common characteristics of most of the ecosystems as follows:

1. The ecosystem is a major structural and functional unit of ecology.
2. The structure of an ecosystem is related to its species diversity in the sense that complex ecosystem have high species diversity.
3. The function of ecosystem is related to energy flow and material cycles within and outside the system.
4. The relative amount of energy needed to maintain an ecosystem depends on its structure. Complex ecosystems needed less energy to maintain them.
5. Young ecosystems develop and change from fewer complexes to more complex ecosystems, through the process called succession.

6. Each ecosystem has its own energy budget, which cannot be exceeded.
7. Adaptation to local environmental conditions is the important feature of the biotic components of an ecosystem, failing which they might perish.
8. The function of every ecosystem involves a series of cycles, e.g., water cycle, nitrogen cycle, oxygen cycle, etc. these cycles are driven by energy. A continuation or existence of ecosystem demands exchange of materials/nutrients to and from the different components.

DYNAMIC OF ECOLOGICAL SYSTEMS

In terrestrial ecosystems, the earlier timing of spring events, and poleward and upward shifts in plant and animal ranges, have been linked with high confidence to recent warming. Future climate change is expected to particularly affect certain ecosystems, including tundra, mangroves, and coral reefs. It is expected that most ecosystems will be affected by higher atmospheric CO_2 levels, combined with higher global temperatures. Overall, it is expected that climate change will result in the extinction of many species and reduced diversity of ecosystems.

Biodiversity

Deforestation on a human scale results in declines in biodiversity and on a natural global scale is known to cause the extinction of many species. The removal or destruction of areas of forest cover has resulted in a degraded environment with reduced biodiversity. Forests support biodiversity, providing habitat for wildlife; moreover, forests foster medicinal conservation. With forest biotopes being irreplaceable source of new drugs (such as taxol), deforestation can destroy genetic variations (such as crop resistance) irretrievably. Since the tropical rainforests are the most diverse ecosystems on Earth and about 80 per cent of the world's known biodiversity could be found in tropical rainforests, removal or destruction of significant areas of forest cover has resulted in a degraded environment with reduced biodiversity.

It has been estimated that we are losing 137 plant, animal and insect species every single day due to rainforest deforestation, which equates to 50,000 species a year. Others state that tropical rainforest deforestation is contributing to the ongoing Holocene mass extinction. The known extinction rates from deforestation rates are very low, approximately 1 species per year from mammals and birds which extrapolates to approximately 23,000 species per year for all species. Predictions have been made that more than 40 per cent of the animal and plant species in Southeast Asia could be wiped out in the 21st century. Such predictions were called into question by 1995 data that show that within regions of Southeast Asia much of the original forest has been converted to monospecific plantations, but that potentially endangered species are few and tree flora remains widespread and stable.

Scientific understanding of the process of extinction is insufficient to accurately make predictions about the impact of deforestation on biodiversity. Most predictions of forestry related biodiversity loss are based on species-area models, with an underlying assumption that as the forest declines species diversity will decline similarly. However, many such models have been proven to be wrong and loss of habitat does not necessarily lead to large scale loss of species. Species-area models are known to overpredict the number of species known to be threatened in areas where actual deforestation is ongoing, and greatly overpredict the number of threatened species that are widespread.

Hydrological

The water cycle is also affected by deforestation. Trees extract groundwater through their roots and release it into the atmosphere. When part of a forest is removed, the trees no longer evaporate away this water, resulting in a much drier climate. Deforestation reduces the content of water in the soil and groundwater as well as atmospheric moisture. The dry soil leads to lower water intake for the trees to extract. Deforestation reduces soil cohesion, so that erosion, flooding and landslides ensue. Shrinking forest cover lessens the landscape's capacity to intercept, retain and transpire precipitation. Instead of trapping precipitation, which then percolates to groundwater systems, deforested areas become sources of surface water runoff, which moves much faster than subsurface flows. That quicker transport of surface water can translate into flash flooding and more localised floods than would occur with the forest cover. Deforestation also contributes to decreased evapotranspiration, which lessens atmospheric moisture which in some cases affects precipitation levels downwind from the deforested area, as water is not recycled to downwind forests, but is lost in runoff and returns directly to the oceans. According to one study, in deforested north and northwest China, the average annual precipitation decreased by one third between the 1950s and the 1980s.

Trees, and plants in general, affect the water cycle significantly:

- their canopies intercept a proportion of precipitation, which is then evaporated back to the atmosphere (canopy interception);
- their litter, stems and trunks slow down surface runoff;
- their roots create macropores – large conduits – in the soil that increase infiltration of water;
- they contribute to terrestrial evaporation and reduce soil moisture via transpiration;

- their litter and other organic residue change soil properties that affect the capacity of soil to store water.
- their leaves control the humidity of the atmosphere by transpiring. 99 per cent of the water absorbed by the roots moves up to the leaves and is transpired.

As a result, the presence or absence of trees can change the quantity of water on the surface, in the soil or groundwater, or in the atmosphere. This in turn changes erosion rates and the availability of water for either ecosystem functions or human services. The forest may have little impact on flooding in the case of large rainfall events, which overwhelm the storage capacity of forest soil if the soils are at or close to saturation. Tropical rainforests produce about 30 per cent of our planet's fresh water.

Soil

Undisturbed forests have a very low rate of soil loss, approximately 2 metric tons per square kilometre (6 short tons per square mile). Deforestation generally increases rates of soil erosion, by increasing the amount of runoff and reducing the protection of the soil from tree litter. This can be an advantage in excessively leached tropical rain forest soils. Forestry operations themselves also increase erosion through the development of roads and the use of mechanised equipment. China's Loess Plateau was cleared of forest millennia ago. Since then it has been eroding, creating dramatic incised valleys, and providing the sediment that gives the Yellow River its yellow colour and that causes the flooding of the river in the lower reaches (hence the river's nickname 'China's sorrow').

Removal of trees does not always increase erosion rates. In certain regions of southwest US, shrubs and trees have been encroaching on grassland. The trees themselves enhance the loss of grass between tree canopies. The bare intercanopy areas become highly erodible. The US Forest Service, in Bandelier National Monument for example, is studying how to restore the former ecosystem, and reduce erosion, by removing the trees. Tree roots bind soil together, and if the soil is sufficiently shallow they act to keep the soil in place by also binding with underlying bedrock. Tree removal on steep slopes with shallow soil thus increases the risk of landslides, which can threaten people living nearby. However most deforestation only affects the trunks of trees, allowing for the roots to stay rooted, negating the landslide.

Forest Transition Theory

The forest area change may follow a pattern suggested by the forest transition (FT) theory, whereby at early stages in its development a country is characterised by high forest cover and low deforestation rates (HFLD countries). Then deforestation rates accelerate (HFHD, high forest cover – high deforestation rate), and forest cover is reduced (LFHD. low forest cover – high deforestation rate), before the deforestation rate slows (LFLD, low forest cover – low deforestation rate), after which forest cover stabilises and eventually starts recovering. FT is not a "law of nature," and the pattern is influenced by national context (for example, human population density, stage of development, structure of the economy), global economic forces, and government policies. A country may reach very low levels of forest cover before it stabilises, or it might through good policies be able to "bridge" the forest transition.

FT depicts a broad trend, and an extrapolation of historical rates therefore tends to underestimate future BAU deforestation for counties at the early stages in the transition (HFLD), while it tends to overestimate BAU deforestation for countries at the later stages (LFHD and LFLD).

Countries with high forest cover can be expected to be at early stages of the FT. GDP per capita captures the stage in a country's economic development, which is linked to the pattern of natural resource use, including forests. The choice of forest cover and GDP per capita also fits well with the two key scenarios in the FT:

(i) a forest scarcity path, where forest scarcity triggers forces (for example, higher prices of forest products) that lead to forest cover stabilisation; and

(ii) an economic development path, where new and better off-farm employment opportunities associated with economic growth (= increasing GDP per capita) reduce profitability of frontier agriculture and slows deforestation.

Historical Causes

Prehistory: The Carboniferous Rainforest Collapse, was an event that occurred 300 million years ago. Climate change devastated tropical rainforests causing the extinction of many plant and animal species. The change was abrupt, specifically, at this time climate became cooler and drier, conditions that are not favourable to the growth of rainforests and much of the biodiversity within them. Rainforests were fragmented forming shrinking 'islands' further and further apart. This sudden collapse affected several large groups, effects on amphibians were particularly devastating, while reptiles fared better, being ecologically adapted to the drier conditions that followed. Rainforests once covered 14 per cent of the earth's land surface; now they cover a mere 6 per cent and experts estimate that the last remaining rainforests could be consumed in less than 40 years. Small scale deforestation was practiced by some societies for tens of thousands of years before the beginnings of civilization. The first evidence of deforestation appears in the Mesolithic period. It was probably used to convert closed forests into more open ecosystems favourable to game animals. With the advent of agriculture, larger areas began to be deforested, and fire became the prime tool to clear land for crops. In Europe there is little solid evidence before 7000 BC. Mesolithic foragers used fire to create openings for red deer and wild boar. In Great Britain, shade-tolerant species such as oak and ash are replaced in the pollen record by hazels, brambles, grasses and nettles. Removal of the forests led to decreased transpiration, resulting in the formation of upland peat bogs. Widespread decrease in elm pollen across Europe between 8400–8300 BC and 7200–7000 BC, starting in southern Europe and gradually moving north to Great Britain, may represent land clearing by fire at the onset of Neolithic agriculture.

The Neolithic period saw extensive deforestation for farming land. Stone axes were being made from about 3000 BC not just from flint, but from a wide variety of hard rocks from across Britain and North America as well. They include the noted Langdale axe industry in the English Lake District, quarries developed at Penmaenmawr in North Wales and numerous other locations. Rough-outs were made locally near the quarries, and some were polished locally to give a fine finish. This step not only increased the mechanical strength of the axe, but also made penetration of wood easier. Flint was still used from sources such as Grimes Graves but from many other mines across Europe. Evidence of deforestation has been found in Minoan Crete; for example the environs of the Palace of Knossos were severely deforested in the Bronze Age.

Pre-industrial History

Throughout most of history, humans were hunter gatherers who hunted within forests. In most areas, such as the Amazon, the tropics, Central America, and the Caribbean, only after shortages of wood and other forest products occur are policies implemented to ensure forest resources are used in a sustainable manner.

In ancient Greece, Tjeered van Andel and co-writers summarised three regional studies of historic erosion and alluviation and found that, wherever adequate evidence exists, a major phase of erosion follows, by about 500-1,000 years the introduction of farming in the various regions of Greece, ranging from the later Neolithic to the Early Bronze Age. The thousand years following the mid-first millennium BCE saw serious, intermittent pulses of soil erosion in numerous places. The historic silting of ports along the southern coasts of Asia Minor (*e.g.* Clarus, and the examples of Ephesus, Priene and Miletus, where harbors had to be abandoned because of the silt deposited by the Meander) and in coastal Syria during the last centuries BC.

Easter Island has suffered from heavy soil erosion in recent centuries, aggravated by agriculture and deforestation. Jared Diamond gives an extensive look into the collapse of the ancient Easter Islanders in his book *Collapse*. The disappearance of the island's trees seems to coincide with a decline of its civilization around the 17th and 18th century. He attributed the collapse to deforestation and over-exploitation of all resources.

The famous silting up of the harbor for Bruges, which moved port commerce to Antwerp, also followed a period of increased settlement growth (and apparently of deforestation) in the upper river basins. In early medieval Riez in upper Provence, alluvial silt from two small rivers raised the riverbeds and widened the floodplain, which slowly buried the Roman settlement in alluvium and gradually moved new construction to higher ground; concurrently the headwater valleys above Riez were being opened to pasturage. A typical progress trap was that cities were often built in a forested area, which would provide wood for some industry (for example, construction, shipbuilding, pottery). When deforestation occurs without proper replanting, however; local wood supplies become difficult to obtain near enough to remain competitive, leading to the city's abandonment, as happened repeatedly in Ancient Asia Minor. Because of fuel needs, mining and metallurgy often led to deforestation and city abandonment.

With most of the population remaining active in (or indirectly dependent on) the agricultural sector, the main pressure in most areas remained land clearing for crop and cattle farming. Enough wild green was usually left standing (and partially used, for example, to collect firewood, timber and fruits, or to graze pigs) for wildlife to remain viable. The elite's (nobility and higher clergy) protection of their own hunting privileges and game often protected significant woodlands.

Major parts in the spread (and thus more durable growth) of the population were played by monastical 'pioneering' (especially by the Benedictine and Commercial orders) and some feudal lords' recruiting farmers to settle (and become tax payers) by offering relatively good legal and fiscal conditions. Even when speculators sought to encourage towns, settlers needed an agricultural belt around or sometimes within defensive walls. When populations were quickly decreased by causes such as the Black Death or devastating warfare (for example, Genghis Khan's Mongol hordes in eastern and central Europe, Thirty Years' War in Germany), this could lead to settlements being abandoned. The land was reclaimed by nature, but the secondary forests usually lacked the original biodiversity.

From 1100 to 1500 AD, significant deforestation took place in Western Europe as a result of the expanding human population. The large-scale building of wooden sailing ships by European (coastal) naval owners since the 15th century for exploration, colonisation, slave trade–and other trade on the high seas consumed many forest resources. Piracy also contributed to the over harvesting of forests, as in Spain. This led to a weakening of the domestic economy after Columbus' discovery of America, as the economy became dependent on colonial activities (plundering, mining, cattle, plantations, trade, etc.)

In *Changes in the Land* (1983), William Cronon analysed and documented 17th-century English colonists' reports of increased seasonal flooding in New England during the period when new settlers initially cleared the forests for agriculture. They believed flooding was linked to widespread forest clearing upstream.

The massive use of charcoal on an industrial scale in Early Modern Europe was a new type of consumption of western forests; even in Stuart England, the relatively primitive production of charcoal has already reached an impressive level. Stuart England was so widely deforested that it depended on the Baltic trade for ship timbers, and looked to the untapped forests of New England to supply the need. Each of Nelson's Royal Navy war ships at Trafalgar (1805) required 6,000 mature oaks for its construction. In France, Colbert planted oak forests to supply the French navy in the future. When the oak plantations matured in the mid-19th century, the masts were no longer required because shipping had changed.

Norman F. Cantor's summary of the effects of late medieval deforestation applies equally well to Early Modern Europe:

> *Europeans had lived in the midst of vast forests throughout the earlier medieval centuries. After 1250 they became so skilled at deforestation that by 1500 they were running short of wood for heating and cooking. They were faced with a nutritional decline because of the elimination of the generous supply of wild game that had inhabited the now-disappearing forests, which throughout medieval times had provided the staple of their carnivorous high-protein diet. By 1500 Europe was on the edge of a fuel and nutritional disaster [from] which it was saved in the sixteenth century only by the burning of soft coal and the cultivation of potatoes and maize.*

ECOLOGICAL SECURITY

Spiraling population and increasing industrialisation are posing a serious challenge to the preservation of the terrestrial and aquatic ecosystems. Environmental protection is the key to ensuring a healthy life for the people. Environmental problems are on the increase and are more pronounced in densely populated cities. Creation of awareness regarding the ecological hazards among the public is absolutely essential. Environmental conservation and abatement of pollution are critical for sustainable development.

Department of Environment

The Department of Environment was created in 1995 as the nodal Department for dealing with environmental management of the State. The Department is entrusted with the implementation of major projects like pollution abatement in the Cauvery, Vaigai and Tamiraparani rivers, pollution abatement in Chennai city waterways, National Lake Conservation Programme and all aspects of environment other than those dealt with by Tamil Nadu Pollution Control Board.

One of the main objectives of the department is to implement Environmental Awareness Programme Wide publicity is being given on World Environment Day, Ozone day and on Bhogi Day to create environmental awareness among the general public. Due to the concerted efforts, the level of air and noise pollution has been brought down to the tune of 20-25% in the last three years. To create environmental awareness among the school and college studies, 1260 eco-clubs have been formed in all the districts of State involving selected educational institutions and NGOs. It is proposed to strengthen these existing eco-clubs. The outstanding NGOs, experts and individuals are honoured with environmental awards in recognition of their excellent contribution in the field of environment.

Tamil Nadu Pollution Control Board

The Tamil Nadu Pollution Control Board enforces the provisions of the Water (Prevention and Control of Pollution) Act, 1974 as amended, the Water (Prevention and Control of Pollution) Cess Act, 1977 as amended, the Air (Prevention and Control of Pollution) Act, 1981 as amended and the relevant provisions/rules of the Environment (Protection) Act, 1986 to prevent, control and abate pollution and for protection of environment. The Board functions with its Head Office at Chennai. There are 25 District Offices and 14 Environmental Laboratories established by the Board.

Monitoring of Industries

The Board has inventorised about 28,000 industries. The Board has prescribed standards for discharge of effluent, ambient air quality and gaseous emissions from various industries and the industries have to take necessary pollution control measures to meet the standards prescribed by the Board. For effective monitoring, the Board has classified the industries into red, orange and green, based on their pollution potential.

Procedure for Issue of Consent

The Board issues consent to industries in two stages under the Water Act and the Air Act for establishment and operation of industrial units. Consent to establish is issued depending upon the suitability of the site, before the industry takes up the construction activity. Consent to operate is issued after installation of effluent treatment plant and air pollution control measures, before commissioning production. Consent is issued subject to general conditions and specific conditions.

Vehicle Emission Monitoring

The Board is carrying out the vehicle emission monitoring in Chennai, Dindigul, Palani, Udhagamandalam and Chengalpattu. In addition, private agencies have been authorised by the Transport Department in Chennai city to check the emission level of the vehicles.

The Board has upgraded and computerised all its vehicle emission monitoring stations for testing diesel driven vehicles. The Transport Corporations have also been instructed to closely monitor the emission levels of their buses.

For controlling vehicular emission, cleaner fuel like unleaded petrol, petrol with 3% benzene and low sulphur fuel (0.05%) have been introduced in Chennai Metropolitan Area. Passenger cars complying with Bharat stage-II norms alone are registered in Chennai since July 2001. 2T oil auto dispensing system have been provided in retail outlets.

The Board is also participating in a research project with an non governmental organisation and the Civil Supplies Department to study the use of gas chromatograph to detect fuel adulteration.

Action has already been taken to introduce auto liquefied petroleum gas in Chennai as it is a cleaner fuel. Steps are being taken to popularise the use of liquefied petroleum gas for autorickshaws, call taxis and other private vehicles which will help in improving air quality.

(1) Vehicle Emission Monitoring Stations provided by the Board. 8 Nos.
(2) Vehicle Emission Monitoring Stations provided by the Board in MTC Depots. 6 Nos.
(3) Emission Checking Stations provided by Private Agencies. 236 Nos.
(4) Auto Liquefied Petroleum Gas Dispensing Stations commissioned in Chennai. 12 Nos.

Noise Level Monitoring

Towards controlling noise pollution in urban areas, about 52,586 air horns were removed as of December 2004 from buses and lorries throughout the State. All the districts have been declared as air horn free districts. For noise level monitoring at the district level, sophisticated noise level meters have been provided to the District Offices of the Board.

Water Quality Monitoring

Pollution of major rivers in the State is caused by the discharge of untreated sewage from the urban local bodies and panchayats and untreated or partially treated effluent from industries. In case of industrial pollution, it is the responsibility of the industrial units to provide the required effluent treatment plants either individually or collectively so as to achieve the standards. Various pollution abatement schemes are being implemented under the National River Conservation Programme under the coordination of the Department of Environment.

Water Quality Monitoring Programmes

Under the Global Environmental Monitoring System, the Board is closely monitoring the quality of water in the Cauvery basin at Mettur, Pallipalayam, Musiri and ground water quality at Musiri. Similarly, water quality of rivers Cauvery (16 stations), Tamiraparani (7 stations), Palar (1 station) and Vaigai (1 station) and the three important lakes in Udhagamandalam, Kodaikkanal and Yercaud are being monitored under the Monitoring of Indian National Aquatic Resources System by the Board. The Board is continuously monitoring the Chennai city water ways to prevent pollution due to discharge of trade effluent from industries and sewage from local bodies and is collecting and analysing samples of river water and outfalls at regular intervals, since 1991.

MAINSTREAMING THE ENVIRONMENT

Global Ecology, International Institutions and the Crisis of Environmental Governance Five years later, at the June 1997 Special Session of the United Nations General Assembly dedicated to the review of UNCED's implementation, the climate was rather different. Optimism had given way to disappointment and, in some cases, there was real concern about the viability of the "sustainable development" model, which relies on a framework of action that does not fully address the causes of environmental destruction. Developed countries have been unable or unwilling to stick to their promise of increasing the aid to development to 0.7% of GDP, as agreed in Rio. Countries like the United States, the largest contributor to global warming, have not shown the will to take effective action that would show a real commitment to reduce their industrial emissions. On the other hand, developing countries refused to take any further steps without the guarantee that substantive financial resources would back them or that at least the commitments taken in Rio would be respected. The New York 1997 Declaration even recognized that the situation of the environment had deteriorated over the intervening five years, hoping modestly that more progress would be achieved by the next summit in 2002. The meager positive results produced by the massive efforts in the field of international cooperation for the environment seem to indicate the contradictory character of this new, global "environmentalism." The purpose of this article is to demonstrate that, while originally being the potential source of a radical and transformative project, environmental concerns were ultimately reframed by the joint action of technocratic environmentalists, the international UN-related establishment and business and industry sectors to become compatible with global development. Adopting an international political economy perspective, the article explores the interaction between state and markets in the construction of global environmental politics. It provides evidence that although there is a new consensus on the diagnosis of the problem - worldwide environmental degradation - very few commitments have been taken to alter the accumulation model and the patterns of production and consumption that contribute to this situation.

It suggests that the failure of the international system in ensuring a move towards sustainability, exemplified in New York, is linked to the very nature of the global bargain struck in Rio. By aiming to make "development" - in its more recent global phase, with its focus on globalized and ever expanding production, trade and consumption - become "sustainable," the concept of sustainability has been stripped of most of its meaning. The inability of the international community to deal with most global environmental issues reveals the contradictory nature of the "sustainable development" consensus and demonstrates the limits of international cooperation in the name of the environment.

Origins and Dimensions of the Ecological Project

In order to understand the meaning of the transformation of environmental concerns into a widely accepted concept, it is useful to recall the original purpose of the ecological project. The ecological movement finds its origins in a protest aimed at defending the right of individuals to regain influence over their ways of living, of producing, and of consuming. As stressed by Gorz (1992), it started as a radical cultural movement, as an attempt by individuals to control and understand the consequences of their actions. With the ecological critique, activists hoped to refocus attention on local knowledge and practices and to bridge the separation of humans from nature, a division that had been at the heart of the Enlightenment project.

In the 1970s, the ecological movement became a political movement, and there was an awareness that the demands of ecology were not only sectorial and local aspirations but rather represented a value shared across national divides (Smith 1996; Gorz 1992).

The publication of the report "Limits to Growth" by the Club of Rome in 1972 gave a scientific backing to these cultural demands and showed the risks posed by the model of industrial growth on the future of life on earth. The report provided a holistic view of the interrelationship between population growth, food production and consumption, the industrialization process, depletion of nonrenewable resources and waste and pollution at the global level, recognizing that waste and pollution are not only a problem for the living conditions and consumption patterns of the population, but affect the very basis of the productive sphere's reproduction.

For the first time, environmental degradation provoked by economic growth was considered from a global perspective, going beyond the occasional questioning of pollution problems during the 1950s and 1960s. In addition, the report launched a real debate on the morality of growth and of the differences in consumption and living standards between developed and developing countries. The 1970s also represented an inflection in the history of social mobilization and collective action with the emergence of the "new social movements," which identify themselves as value movements carrying universal interests going beyond class, nation, sex and race borders. The new social movements such as the environmental movement appear as "modern" in the sense that they are based upon the belief that history's course can be changed by social actors and are not determined by what Touraine calls a "metasocial principle" (Offe 1988, 219). Environmentalists believe that, although representing a real challenge to our present lifestyles and habits, it is possible to move towards a sustainable society that respects nature and privileges well-being over accumulation. Speaking about the existence of a unique and unified "green movement" is clearly incorrect. Environmental concerns mean different things to different people, take many forms and are expressed through different channels. In addition, environmentalism takes very different forms in developed or in developing countries. It can mean fighting for an even better quality of life in advanced countries, and fighting for subsistence or even survival in poor countries. Despite this diversity, for the purpose of academic inquiry, three main components of the "green movement," albeit sometimes overlapping, can be distinguished. These three categories should be viewed as "ideal-typical" and not necessarily mutually exclusive.

The first tendency of the ecological movement, deep ecology, is typically a postmodern movement.

In philosophical terms, deep ecology challenges the separation between humans and nature that was at the heart of modern humanism. Deep ecology is not "anthropocentric," it is "ecocentric." As observed by Merchant (1992), it seeks a total transformation in science and in worldviews that will lead to the replacement of the mechanistic paradigm (which has dominated the past three hundred years) by an ecological framework of interconnectedness and reciprocity. The ideas of deep ecology have influenced (among others) Greenpeace, the largest green NGO, which claims that humanist value systems must be replaced by supra-humanist values that place any vegetal or animal life in the sphere of legal and moral consideration. Greenpeace is therefore an example of an environmental organization which, based on scientific reports and examinations, acts to change worldviews and consciousness in order to promote a shift to "ecocentrism" rather than trying to act to transform the production systems which lie at the root of environmental problems.

Yet, while having influenced the most well-known environmental NGO, deep ecology remains a fairly marginal wing of the green movement. Deep ecologists have been criticized for their lack of a political critique, failing to recognize that the idea itself of "ecocentrism" is "anthropocentric." As stressed by Merchant, deep ecologists take the character of capitalist democracy for granted rather than submitting it to a critique. Their tendency to refuse to consider economic policy and to assume a purely conservationist standpoint relegates them to a secondary position.

The second component of the "green movement" is what can be called the "social ecology" movement, which is to a large extent composed of people from the "New Left," dissatisfied with Marxism. Contrary to the deep ecologists, social ecologists maintain an anthropocentric perspective: the concern for nature is understood as a concern for the environment of human beings. Social ecologists seek transformations in production and reproduction systems, that is, a transformation of political economy, as the way to achieve sustainability, social equity and well being. Social ecologists see a contradiction between the logic of capitalism and the logic of environmental protection. For them, environmental protection cannot be made dependent upon economic development, because development, in its liberal sense, has meant the subordination of every aspect of social life to the market economy, and can therefore no longer be considered as a desirable goal. The hegemonic view on "sustainable development," which rehabilitates development as the global goal of humans, is thus unsatisfactory. Social ecologists call for a rethinking of the theoretical basis of development that should include not only economic but also political and epistemological dimensions, such as the questions of participation, of empowerment and local knowledge systems.

For them, what makes development "unsustainable" at the global level is the pattern of consumption in rich countries. Thinking about sustainability thus implies considering the contradictions imposed by the structural inequalities of the global system. Finally, social ecologists vary to a certain extent in the North and in the South: generally speaking, organizations in the North sometimes carry their rejection of development as far as to strike postmodern stances, while organizations in the South focus more on equity and on the need to redistribute the benefits of development.

Finally, there is a more technocratic tendency to the green movement, a tendency that tries to make economic growth and environmental protection appear as compatible goals, which need not require a profound change in values, motivations and economic interests of social actors, nor new models of economic accumulation. For them, it is because capitalist production methods and life standards are not developed enough that environmental problems emerge.

The evidence is that environmental standards are higher in richer countries. Technocratic environmentalists seek to preserve the environment through the establishment of international institutions, the use of economic and market instruments and the development of clean and "green" technology. The result is a rather apolitical approach and activists who, though still interested in environmental protection, are not primarily committed to ideas of equity and social justice, or at least not as committed as social ecologists (Gudynas 1993). The technocratic tendency is thus essentially a rich country tendency, although it is also present in some elite circles in the South.

These environmentalists tend to focus on issues of population for example, arguing that the biggest threat to the environment comes from high population growth in the Third World and the pressure it will bring to bear on the stock of natural resources. Technocratic environmentalists usually tend to belong to organizations which have little or no membership, and rely on their technical and legal expertise and on their research and publishing programs to influence decision-making. Through their close relationship with government and other influential actors and their easy access to international organizations, these organizations tend to have a greater impact than activist membership organizations.

Today, it can be said that this technocratic approach appears to be prevailing over both the biocentric (deep ecology) and the social ecology perspectives and has become what is today mainstream environmentalism, which finds its major expression in the concept of "sustainable development." Despite the challenging and radical nature of ecological concerns, the fact that they might present a potential for change in the present economic model, they were ultimately reframed so as to constitute what appears as an apolitical, techno-managerial approach.

The Formation of a Consensus on "Sustainable Development" It is interesting to examine how the apparent consensus around the concept of "sustainable development" was built and how the project of global environmental "management" became hegemonic. Two main actors have contributed to the hegemony of the liberal environmental management project.

One is the scientific and policy-making environmental community, or, in the words of Peter Haas, the environmental "epistemic community" (Haas 1990); the other actor is business and industry.

The Brundtland Report, the United Nations Conference and the Global North-South "Bargain" International environmental politics did not emerge in the 1990s. As early as 1972, a United Nations Conference on the Human Environment took place in Stockholm, launching the era of international environmental negotiations.

Stockholm did produce some significant outcomes, leading to the creation of the United Nations Environment Program (UNEP), based in Nairobi, which coordinates environmental action within the United Nations. The context of the Stockholm Conference was not very favorable to the adoption of strong environmental commitments. Developing countries were unsatisfied with the UN system and preparing the movement for a New International Economic Order.

They were not willing to yield part of their sovereignty over natural resources in the name of environmental protection, and denounced the emergence of "eco-imperialism." The oil crisis of the 1970s relegated environmental protection to a marginal position in international relations. In the 1980s, the international climate started to change as the debt crisis was seriously affecting developing countries and their role and participation in international fora. In this context, "international commissions" were established to try to elaborate global proposals to promote peace and development, such as the Brandt Commission. Efforts were also undertaken to replace environmental protection on the international political agenda. The World Commission on Environment and Development was established in 1983 under the presidency of Gro Harlem Brundtland, and asked to produce a comprehensive report on the situation of the environment at the global level.

The work of the Commission represented a landmark in international initiatives to promote environmental protection as it produced the concept of sustainable development, a concept that would become the basis of environmental politics worldwide. Sustainable development is defined by the Brundtland Report as a development that is "consistent with future as well as present needs" (World Commission on Environment and Development 1987).

The concept of sustainable development was built as a political expression of the recognition of the "finiteness" of natural resources and of its potential impact on economic activities. Indeed, the report argues that, while we have in the past been concerned about the impacts of economic growth upon the environment, we are now forced to concern ourselves with the impacts of ecological stress - degradation of soils, water regimes, atmosphere and forests-upon our economic prospects. The report offered a holistic, global vision of today's situation by arguing that the environmental crisis, the developmental crisis and the energetic crisis are all part of the same, global crisis. It offers solutions to this global crisis, which are mainly of two kinds. On the one hand there are solutions based on international cooperation, with the aim of achieving an international economic system committed to growth and the elimination of poverty in the world, able to manage common goods and to provide peace, security, development and environmental protection. On the other hand, come recommendations aiming at institutional and legal change, including measures not only at the domestic level but also at the level of international institutions. The report emphasizes the expansion and improvement of the growth-oriented industrial model of development as the way to solve the global crisis.

The Brundtland Report also promoted the view that global environmental degradation can be seen as a source of economic disruption and political tension, therefore entering the sphere of strategic considerations. For the Brundtland Commission, the traditional forms of national sovereignty are increasingly challenged by the realities of ecological and economic interdependence, especially in the case of shared ecosystems and of "global commons," those parts of the planet that fall outside national jurisdictions. Here, sustainable development can be secured only through international cooperation and agreed regimes for surveillance, development, and management on the common interest.

For example, the consequences of climate change such as rising sea levels and the effects of temperature variations on agricultural production would require deep changes in the economy and impose high costs on all countries, thus leading to very unstable situations. The issue of forest preservation can also fit into this context, since forests contribute to the stability of climate by acting as carbon sinks, and assure the regeneration of ecosystems by providing reservoirs of biological diversity. Preserving forests then becomes more than an ecological concern: it is also a security imperative. So the "environmental security" discourse was also a cause for the need to find a "consensual solution" to issues of environmental protection. The United Nations Conference on Environment and Development (UNCED), held in Rio de Janeiro in June 1992, marked the official institutionalization of environmental issues in the international political agenda.

Twenty years after the 1972 Stockholm Conference, which was on the "Human Environment," Rio meant a real shift in the vision that had dominated environmental politics so far. After Rio, environmental considerations became incorporated into development, and a "global bargain" was struck between North and South on the basis of the acceptance from both sides of the desirability of achieving a truly global economy which would guarantee growth and better environmental records to all. UNCED recognized the "global finiteness" of the world, i.e., the scarcity of natural resources available for development, but adopted the view that, if the planet is to be saved, it will be through more and better development, through environmental management and "eco-efficiency." The UNCED process involved over a hundred and fifty hours of official negotiations spread over two and a half years, including two planning meetings, four Preparatory Committees (Prepcoms), and the final negotiation session at the Rio Summit in June 1992.

The major result of UNCED is called "Agenda 21," a 700-page global plan of action which should guide countries towards sustainability through the 21st century, encompassing virtually every sector affecting environment and development. Besides Agenda 21, UNCED produced two non-binding documents, the "Rio Declaration" and the Forest Principles. In addition, the climate change and the biodiversity conventions, which were negotiated independently of the UNCED process in different fora, were opened for signature during the Rio Summit and are considered as UNCED-related agreements. The "Rio Declaration," which was the subject of much dispute between the Group of 77 (the coalition of developing countries) and industrialized countries, mainly the United States, illustrates well the kind of bargain reached in Rio.

It recognizes the "right of all nations to development" and their sovereignty over their national resources, identifies "common but differentiated responsibility" for the global environment, and emphasizes the need to eradicate poverty, all demands put forward by the Group of 77. In return, the suggestions by the G77 to include consumption patterns in developed countries as the "main cause" of environmental degradation and the call for "new and additional resources and technology transfer on preferential and concessional terms" were rejected by OECD countries.

In the end, on the issue of finance, an institution called the "Global Environment Facility" (GEF) was set, under the joint administration of the World Bank, the United Nations Development Program (UNDP) and the United Nations Environment Program (UNEP), as the only funding mechanism on global environmental issues, and OECD countries committed themselves to achieving a target of 0.7 percent of GNP going to ODA (Overseas Development Assistance) by the year 2000, to help developing countries implement UNCED's decisions.

Despite the failure of the G77 to win significant concessions on financial resources, if one considers the differences in priorities between developed and developing countries and the conflictual character of the negotiation process, UNCED's outcomes were still seen by the international establishment as quite impressive, marking "an important new stage in the longer-term development of national and international norms and institutions needed to meet the challenge of environmentally sustainable development."

A Commission on Sustainable Development (CSD) was established to monitor and report on progress towards implementing UNCED's decisions. In particular, the CSD's stated aims are to enhance international cooperation by rationalizing the intergovernmental decision-making capacity, and to examine progress in the implementation of Agenda 21 at the national, regional and international levels.

After UNCED, environmental considerations were "integrated" at all levels of action. The "sustainable development paradigm," as some authors recognize, is already replacing the "exclusionist paradigm" (i.e., the idea of an infinite supply of natural resources) in some multilateral financial institutions, as well as in some state bureaucracies and in some parliamentary committees. Most economists now acknowledge that natural resources are scarce and have a value that should be internalized in costs and prices.

Organizations such as the European Union made the "integration" of environmental concerns one of their leading policy principles.

Many countries carried out environmental policy reform to implement UNCED's decisions and the Agenda 21. The boundaries of environmental politics were broadened and its links with all other major issues on the international arena, such as trade, investments, debt, transports, for example, were examined. Efforts were also undertaken to improve environmental records of multilateral finance and development institutions. The World Bank, which has a long history of contributing to environmental degradation by financing destructive projects, went through a "greening" process, and now has a "Department of the Environment" which conducts "environmental impact assessments" and imposes "environmental conditionalities" before granting loans. The World Trade Organization has a "Committee on Trade and Environment" (CTE) which is in charge of ensuring that open trade and environmental protection are mutually supportive. All these efforts can be seen, according to Porter and Brown (1996), as part of a longer-term process of evolution toward environmentally sound norms governing trade, finance, management of global commons, and even domestic development patterns. Environmental considerations were then to be introduced in all major international bureaucracies as a dimension to take into consideration in decision-making processes, and as a challenge for global management. To a certain extent, the "technocratic" approach became hegemonic because it best suited the interests of the international development elite as it magnified its managerial responsibilities. In a time when the legitimacy and utility of the United Nations system was being seriously questioned by its idealizer and major financial supporter - the United States - the goal of making environment and development compatible was seized by some UN agencies as an unexpected opportunity to regain credibility, as well as to be granted funds and to hire new staff for recently created units on "trade and environment" or "finance and environment." UNCED provided a new legitimacy to international organizations such as the World Bank or the World Trade Organization and to their bureaucracies, which now try to assume a leading role in "managing the earth." With the promotion of economic growth to a planetary imperative and the rehabilitation of technological progress, both development institutions and organizations and states appeared as legitimate agents to solve global environmental problems.

If international organizations have benefited from the global perspective that emerged from Rio, they have also contributed to mold it. There is an active "epistemic community," which includes both the international organization establishment and large environmental NGOs, promoting the "global environmental management" approach.

These groups tend to believe that their moral views are cosmopolitan and universal, and emphasize the existence of an international society of human beings sharing common moral bonds.

In this kind of "same boat" ideology, environmental concerns tend to be presented as moral imperatives, related neither to political nor to economic advantages. It would be a consensual concern, a sort of universal principle accepted over borders and political boundaries. An example of an institution promoting these ideas is given by the Commission on Global Governance. In the words of the Commission, "we believe that a global civic ethic to guide action within the global neighborhood and leadership infused with that ethic are vital to the quality of global governance. We call for a common commitment to core values that all humanity could uphold. We further believe humanity as a whole will be best served by recognition of a set of common rights and responsibilities."

Part of the Green movement came to support this "same boat ideology" and was incorporated into the epistemic community. Actually, mainstream conservationist environmentalists were fully admitted into the global environmental management establishment, conferring legitimacy to the UNCED process.

NGOs contributed to UNCED to a degree unprecedented in the history of UN negotiations. NGOs lobbied at the official process, participated in Prepcoms and were even admitted in some countries' delegations, a novelty which was rendered possible by resolution 44/228 calling for "relevant non-governmental organizations in consultative status with the Economic and Social Council to contribute to the Conference, as appropriate."

In addition, during UNCED, NGOs organized in Rio a meeting which ran parallel to the official governmental conference. The "Global Forum," which gathered about 30,000 people, represented 760 associations, among participants and visitors, in a sort of "NGO city." During one week, the Global Forum became home to environmentalists and social activists, to Indians and ethnic minorities, and to feminists and homosexual groups, all united to "save the earth." NGOs organized many demonstrations protesting against the modest results of the official summit and elaborated their own agenda for improving environmental protection worldwide.

Yet, in the eyes of some observers, NGO efforts tended to become coopted by larger and richer groups from advanced countries, which had more means, not only financially but also in terms of organizational, scientific and research capac- ity, to promote their own views (Chatterjee and Finger 1994).

In the end, NGOs decided that they would sign, in Rio, NGOs "treaties" on all the issues being discussed at the UNCED official meeting. The main activity at the Global Forum was then the "treaty negotiation" process, just like at the official forum, a process which proved to be very disappointing, as the same North-South conflicts that were blocking UNCED tended to separate northern and southern NGOs.

Ultimately, the NGO treaty process was little more than a pantomime of real diplomacy, and ultimately, the treaties agreed upon, negotiated among a couple of dozen NGOs, had a very modest impact on the future of NGO activities.

The representation at the Global Forum was also very unequal, illustrating differences in means between northern NGOs, very present, and southern NGOs. Asian, and above all, African NGOs, were severely under-represented. Differences in associative traditions and language barriers also explain the hegemony of Anglo-Saxon organizations at the Global Forum. In the end, influential NGOs decided to concentrate their efforts on lobbying the official conference. The Earth Summit in 1992 thus represented a real moment of acceleration for NGO activities, as it allowed some of them to have a better idea of what their counterparts were doing in other parts of the world, and was the base for establishing cooperation projects and partnerships among organizations. Yet while NGO efforts illustrated by the Global Forum aimed at uniting NGOs worldwide, the green movement came out of Rio appearing even weaker and more fragmented, with the polarization between "realist," co-operative NGOs on the one side and "radical," transformative NGOs on the other.

Finally, the "sustainable development" approach also suited the interests of some governments in the Third World which are primarily committed to economic development and sought through UNCED to obtain concessions in financial and technological terms in exchange of their support for environmental management. Some Third World countries are still marked by a "developmentalist" ideology in which economic development comes before all else.

In addition, resource rich countries such as Malaysia, Indonesia, or Brazil, have traditionally had a vision of unending and expanding frontiers, in which land and natural resources are unlimited and no constraints are seen to exist on the use of resources. As a result, they were unwilling to accept the elaboration of international regimes aiming at limiting their sovereignty over the exploitation of natural resources.

The issue of sovereignty had long been a major source of tension during international environmental negotiations. As long ago as the Stockholm Conference in 1972 developing countries had pressed for the inclusion of a specific principle on the topic. Principle 21 of the Stockholm Declaration stated that "States have, in accordance with the Charter of the United Nations and the principles of international law, the sovereign right to exploit their own resources pursuant to their own environmental policies, and the responsibility to ensure that activities within their jurisdiction or control do not cause damage to the environment of other States or areas beyond the limits of national jurisdiction."

The same debate arose when UNCED was convened, and in the end the sovereignty principle as in stood in the Stockholm Declaration's Principle 21 was included in the Rio Declaration.

In addition, a guarantee that economic development would continue to be the priority on the international agenda was an essential element for developing countries. The reaffirmation of the right to development, and of the sovereignty principle, ensured in Rio, were then the two elements that made agreement at UNCED possible for the Group of 77. The alliance between environment and development could then become official. As described by the vice-president of the International Institute for Environment and Development (IIED), "it has not been too difficult to push the environment lobby of the North and the development lobby of the South together. And there is now in fact a blurring of the distinction between the two, so they are coming to have a common consensus around the theme of Sustainable Development" (World Commission on Environment and Development 1987, 64). Yet to fully understand the nature of this consensus around sustainable development, one last actor needs to be introduced. The actor whose vision shaped most fundamentally the content of this consensus and the real winner of Rio, the business and industry sector, and in particular transnational corporations.

ECOLOGY AND THE POLITICS OF KNOWLEDGE

The paradigm of modern science has evolved in the last few centuries in an environment where all economic activities were aimed at maximising the productivity of man-made processes in individual sectors of the economy. This led to the development of modern technologies with highly negative externalities which remained invisible within the conceptual framework of modern science and economics. This shortcoming emanates from three basic fallacies of modern scientific knowledge:

1. It identifies development merely with sectoral growth, ignoring the underdevelopment introduced in related sectors through negative externalities and the related undermining of the productivity of the ecosystem.
2. It identifies economic value merely with exchange value of marketable resources. ignoring use values of more vital resources and ecological processes.
3. It identifies utilization merely with extraction. ignoring the productive and economic functions of conserved resources.

Development planning based on these false identifications tends to create severe ecological problems because of its inability to recognise ecosystem linkages and the ecological processes operative in the natural world. The ecological relationships between the sectors of natural resources contribute to essential ecological processes which are frequently found to be vital for human survival. Thus, the stability of ecological processes is not merely a matter of aesthetics. An incomplete understanding of the material and economic values of ecological processes leads to the destruction of the material conditions for economic development and eventually survival.

Since the availability of essential and vital resources for survival is dependent on the maintenance of essential ecological processes, economic activities which generate sectoral growth in the shortterm by destroying the essential ecological processes cannot lead to development in the long run. On the contrary, by decreasing the productivity and availability of vital resources, they initiate the process of underdevelopment.

When the natural world is viewed ecologically as a system of interrelated resources which maintain the material basis for human sustenance, economic values can no longer be perceived merely as exchange values in the market. Economic values in the ecological perspective are not always equivalent to their exchange value in the market, evaluated without any significance to their use value.

As a corollary, natural resources can have economic utility that cannot be quantified through the exchange value in the market. Such economic utility includes the maintenance of essential ecological processes that support human survival and, thus, all economic activities. The economic utilisation of resources through extraction may, under certain conditions, undermine and destroy vital ecological processes leading to heavy but hidden diseconomies. The nature of these diseconomies can be understood only through the understanding of ecological processes operating in nature.

The economics of sustenance and basic needs satisfaction is, therefore, linked with ecological perceptions of nature. The economics of sectoral growth on the other hand is related to reductionist science and resource wasteful technologies which are productive in the narrow context of sectoral and labour inputs, but may be counter-productive in the context of the overall economic base of natural resources.

The case studies in the following chapters are only representative of thousands of such cases seen everywhere. They reveal a certain pattern of contemporary economic development which can be identified thus:

1. Development has been equated only with the growth of manufacture in individual industrial sectors and with the increase in productivity of only man-made processes.
2. This sectoral growth of man-made processes has also led to ecological destruction of the natural resource base, affecting negatively other sectors of the economic system. This leads to the decay of systems productivity of all productive processes, man-made and natural.

As a result of this limitation of contemporary economics, economic development has, consequently, been taken to be synonymous with growth. The higher the rate of sectoral growth, the higher is the index of economic development. Possible ecological destruction caused by the resource intensity of sectoral growth that is guided purely by non-ecological economic considerations, has never been introduced in the processes of planning for economic development. The benefit-cost analysis of development projects has thus externalized those ecological changes and is incomplete in three important ways:

1. It deals with benefits and costs as profits and losses in financial ferms.
2. It deals with benefits and costs only in the narrow sectoral perspective and ignores costs generated by inter-sectoral linkages.
3. It deals with benefits that are largely available to more visible and economically powerful groups and ignores costs that are borne by the less visible and economically weaker groups. These costs and the associated underdevelopment are thus made invisible in modern economic analysis.

The utilisation and management of natural resources in India has so far been guided by the narrow and sectoral concept of productivity and restricted benefit-cost analysis. This narrow concept of productivity and benefit-cost analysis has blocked the conceptualization of the criteria of rationality of technology choice which maximizes needs satisfaction while minimising resource use, thus maximising systems productivity.

For example, the clear felling of natural forests in the catchments of rivers, and planting of industrial species of trees has been justified on the grounds of increasing productivity of forests. This concept of productivity is, however, only related to productivity of industrial timber, while forests produce other forms of biomass, like fodder and green mulch, or maintain productivity of soil and water resources. The direct impact of the clear felling of catchment forests on agricultural production through its destructive impact on soil and destabilisation of the hydrological balance is not taken into account in the calculations of the benefits and costs associated with forests.

Regular floods and droughts, which are the consequences of irrational land and water management, are branded as natural disasters for which the whole nation pays heavily. Consequently, the poor and marginal groups which depend on agriculture for their livelihood face increasing impoverishment and poverty. This thrusting of negative externalities on the poor and marginal groups directly leads to the polarisation of society into two groups. One group gains from the process of narrow sectoral growth, while the poor and marginalised majority suffer because of the ecological destruction of natural resources on which they depend for survival.

The dialectical contradiction between the role of natural resources in production processes to generate growth and profits and their role in natural processes to generate stability is made visible by movements based on the politics of ecology. These movements reveal that the perception, knowledge and value of natural resources vary for different interest groups in society.

The politics of ecology is thus intimately linked with the politics of knowledge. For subsistence farmers and forest dwellers a forest has the basic economic function of soil and water conservation, energy and food supplies, etc. For industries the same forest has only the function of being a mine of raw materials. These conflicting uses of natural resources, based on their diverse functions, are dialectically related to conflicting perceptions and knowledge about natural resources. The knowledge of forestry developed by forest dwelling communities therefore evolves in response to the economic functions valued by them. In contrast, the knowledge of forestry developed by forest bureaucracies, which respond largely to industrial requirements, will be predominantly guided by the economic value of maximising raw material production.

The way nature is perceived is therefore related to the pattern of utilisation of resources. Modern scientific disciplines which provide the currently dominant perspectives of nature have generally been viewed es 'objective', 'neutral' and 'universally valid'. These disciplines are, however, particular responses to particular economic interests. This economic determination influences the content and structure of knowledge about natural resources which, in turn, reinforces particular forms of resource utilisation The economic and political values of resource use are thus built into the structure of natural science knowledge.

TECHNOLOGY CHOICE TOWARDS HOLISTIC ECOLOGICAL CRITERIA

When economic development programmes are viewed from the perspective of all the three economies, a clearer view of the political economy of conflicts over natural resources is expected to emerge.

In the dominant mode of economic development, perceived within the framework of the market economy, mediation of technology is assumed to lead to the control of larger and larger quantities of natural resources, thus turning scarcity into abundance and poverty into affluence: Technology, accordingly is viewed as the motive force for development and the vital instrument that guarantees freedom from dependence on nature ' The affluence of the industrialized west is assumed to be associated exclusively with this capacity of modern technology to generate wealth.

The concept of technology per se as a source of abundance and freedom from nature's ecological limits are based in part on the limitations of the market economy in understanding in a holistic manner, the same resources which it exploits. Only when development processes are viewed in the holistic perspective of all the three economies can the scarcities and underdevelopment associated with abundance and development be clearly seen. Most resource-intensive technologies operate in the enclaves with enormous amounts of various resources coming from diverse ecosystems which are normally far away. This long, indirect and spatially distributed process of resource transfer made possible by energy-intensive long distance transportation, leaves invisible the real material demands of the technological processes of development.

The spatial separation of resource exhaustion and the creation of products have also considerably shielded the inequality creating tendencies of modern technologies. Further, it is simply assumed that the benefits of economic development based on these modern technologies will automatically percolate to the poor and the needy and growth will ultimately take care of the problems of distributive justice. This would, of course, be the case, if growth and surplus were in a sense absolute and purchasing power existed in all socio-economic groups. None, however, is correct. Surplus is often generated at the cost of the ecological productivity of natural resources or at the cost of exhausting the capital of non-renewable resources. For the poor, the only impact of such economic activity often is the loss of their resource base for survival.

It is thus no accident that modern, efficient and 'productive' technologies 'creased within the context of growth in market economic terms are associated with heavy social and ecological costs. The resource and energy intensity of the production processes they give rise to demands ever increasing resource withdrawals from the natural ecosystems. These excessive withdrawals in the course of time disrupt essential ecological processes and result in the conversion of renewable resources into non-renewable ones. Over time, a forest provides inexhaustible supplies of water and biomass including wood, if its capital stock, diversity and hydrological stability are maintained and it. is harvested on a sustained yield basis.

The heavy and uncontrolled market demand for industrial and commercial wood, however, requires continuous over-felling of trees which destroys the regenerative capacity of the forest ecosystems and over time converts these forests into non-renewable resources. Sometimes the damage to nature's intrinsic regenerative capacity is impaired not directly by over-exploitation of a particular resource but indirectly by damage caused to other natural resources related through ecological processes. Thus under tropical monsoon conditions, over-felling of trees in catchment areas of streams and rivers not only destroys forest resources, but also stable, renewable sources of water. Resource-intensive industries do not merely disrupt essential ecological processes by their excessive demands for raw materials; they also destroy and disrupt vital ecological processes by polluting essential resources like air and water. In the words of Rothman: 'the private economic rationality of the profit seeking business enterprise is a murderous providence because it cannot guarantee the optimum use of resources for society as a whole. It cannot avoid continually creating situations which cause the pollution of an environment

In the context of resource scarcity where most resources are already being utilised for the satisfaction of survival needs, further diversion of resources to new uses is likely to threaten survival and generate conflicts between the demands of economic growth and the requirements of survival. It, therefore, becomes essential to evaluate the role of new technologies in economic development on the basis of their resource demands and conflict with the demands of survival. The productivity of 8 technology in the perspective of human survival must distinguish outputs in terms of their potential for satisfaction of vital or non-vital needs, because on the continued satisfaction of vital needs depends human survival. As Georgescu-Roegen points out.

There can be no doubt about it. Any use of the natural resources for the satisfaction of non-vital needs means a smaller quantity of life in the future.

If we understand well the problem, the best use of our iron resources is to produce plows or harrows as they are needed, not Rolls Royces, not even agricultural tractors.

In the context of the market economy, the indicators of technological efficiency and productivity are totally independent of the difference between the satisfaction of basic needs and luxury requirements. between resources extracted by ecologically sensitive or insensitive technologies or of the nature of the contribution of economic growth to diverse socio-economic categories. In the context of a highly non-uniform distribution of purchasing power and scanty knowledge of or respect for ecological processes, economic growth depends on production and consumption of nonvital products. The expansion of the formal sector of the economy for the production of non-vital goods often leads to further diversion of vital natural resources. For example, water-intensive production of flowers or fruits for the lucrative export market often results in water scarcity in low rainfall areas. In a world with a limited and shrinking resource base, and in the economic framework of a market economy, non-vital luxury needs are fulfilled at the cost of vital survival needs. The high powered pull of the purchasing capacity of the rich of the world can draw out necessary resources in spite of resource scarcity and resulting conflicts.

This complete lack of recognition of the resource needs of the survival economy nature's economy in the current paradigm of development economics shrouds the political issues arising from resource transfer and ecological destruction. For the economic sector based on 'efficient modern technologies', this provides an ideological weapon for increased control of the sponsors of economic development over the total natural resource endowments of the countries concerned.

The ideological and limited concept of 'productivity' of technologies has been universalised with the consequence that all other costs of the economic process become invisible. The invisible forces which contribute to the increased 'productivity' of a modern farmer or factory worker emanate from the increased consumption of non-renewable natural resources. Lovins has described this as the amount of 'slave' labour at present at work in the world. According to him, each person on earth, on an average, possesses the equivalent of about fifty slaves, each working forty hours a week. Man's annual global energy conversion from all sources (wood, fossil fuel, hydroelectric power, nuclear) at present is approximately 8 x 10 (12) watts. This is more than twenty times the energy content of the food necessary to feed the present world population at the FAO standard per capita requirement of 3,600 cals per day.

In terms of workforce, therefore, the population of the earth is not 4 billion but about 200 billion, the important point being that about 98 per cent of them do not eat conventional food. The inequalities in the distribution of this 'slave' labour between different countries is enormous, the average inhabitant of the USA, for example, having 250 times as many 'slaves' as the 'average Nigerian'. And this, substantially is the reason for the difference in efficiency between the American and Nigerian economies: it is not due to the differences in the average 'efficiency' of the people themselves. There seems no way of discovering the relative efficiencies of Americans and Nigerians: If Americans were short of 249 of every 250'slaves' they possess, who can say how 'efficient' they would prove themselves to be.

The increase in the levels of resource consumption is taken universally as an indicator of economic development. If the present level of resource consumption in the USA is accepted as the development objectives of India, the total resource demands of 'developed' India can be calculated by multiplying the current resource consumption by a factor of 250. Neither our forests nor our fields or rivers can sustain such a 'development'. When per capita resource consumption is considered, the Malthusian argument relating population with resource scarcity does not hold good. More significant than the population factor is the total resource factor. Thus, although many countries of the South have a much larger population than those of the North, the industrialized of the world consumes more grain than all the other three-quarters put together. This high consumption is due to the fact that intensive livestock production in industrialized countries accounts for 67 per cent of their total grain consumption. This Efficient' process of livestock management for the production of meat, as reported by Odium requires 10 calories of energy input to produce one calorie of food energy.

The energy subsidy provided by the capital stock of the earth's non-renewable resources makes a resource inefficient process appear as efficient in the market economy. It is interesting to note that even in the West, nearly a century ago one calorie of food was produced by using a fraction of a calorie of energy input. The same is true in the economics of water resources use in modern agriculture. When the production of high yielding varieties of seeds is evaluated, not on productivity per unit land (tons/ha) but per unit volume of water input (tons/le lit), these miracle seeds of the Green Revolution are seen as two to three times less efficient in food production than, say, the millets. The results of evaluation of the technological efficiency of processes associated with economic development, when reexamined on a holistic basis and optimised against all resource inputs, would generally lead to the conclusion that: 'the much talked of efficiency of widely practiced high technology is not intrinsically true. They are, in fact, highly wasteful of materials and pollutive (that is, destructive to the productive potential of the environment)'.

New technologies in the market economy are innovated for profit maximization and not to encourage resource prudence per se. The extent of inefficiency in the utilisation of natural resources with production processes based on resource-intensive technologies, can be illustrated with the production of soda ash, an important industrial material. In the Solvay process for the production of soda ash. the two materials used are sodium chloride and limestone.

The entire limestone used in the process ends up as waste material, 25 per cent of the sodium chloride is lost as unreacted salt. From the balance 75-80 per cent, the acidic half is lost and only the basic half goes into the final product. Therefore only 40 per cent of the raw materials consumed are actually utilised. The waste products pollute land and water resources systems. The economy of the process is artificially made good by concessions in procuring limestone, salt and fuel and further concessions in respect of land, transport, etc. It is these subsidies for natural resources which make the counter-productive processes appear efficient.

Referring to the technology of production of frozen orange juice Schnaiberg made the following remarks:

What is true of the unobtrusive shift from fresh oranges to frozen orange juice is typical of most transitions from traditional to late industrial technologies. The majority of these become more energy intensive: the energy content of all the necessary production processes increases per unit produced.... The hall mark of modern technology is its typical labour saving quality-not its energy saving aspect."

Guided by a narrow and distorted concept of efficiency and supported by all types of subsidies, technological change in market economy-oriented development continues in the direction of resource intensity, labour displacement and ecological destruction.

The long-term continuation of such processes will lead to the destruction of the resource base of the survival economy and to human labour being rendered dispensable in the production processes of the market economy. The partisan assumptions of modern economic development which cannot internalise the economy of natural processes and the survival economy are thus being raised to the level of universality. As a result, with the expansion of economic development in Third World countries, the resource-intensive and socially partial development is leading to social instability and conflicts. While ecology movements in the industrially advanced countries are directed against more recent threats to survival like pollution, ecology movements in Third World countries have a much longer history related to resource exhaustion and ecological degradation of natural ecosystems. It is in these countries that the holistic ecological criteria for technology choice is needed most urgently.

The process of transformation and utilisation of natural resources for the satisfaction of societal needs determines the economic organization of human societies. At various stages of development, the dominant patterns of utilisation of natural resources have been guided by the dominant pattern of scientific knowledge, and through the generation and use of technologies that actually bridge the gap between natural resources and human needs and requirements.

A special characteristic of human societies is that they can make deliberate choices between different ways of using resources and satisfying needs. The existence of plurality of alternatives in resource use for economic development creates the need for a selection criteria to make rational decisions about the use of natural resources and technological change. A dialectical relationship exists between the criteria of technology choice and the nature of science and technology developed in response to the criteria. Traditional societies as well as modern scientific-industrial societies have adopted different systems of science and technology which differ primarily in the criteria of choice or rationality that guides resource use patterns for human needs satisfaction. The characterization of certain societies as primitive and unscientific is, thus, sociologically and epistemologically unfounded. The fact that values and rationality criteria of one form of social organization generate a particular type of science and technology matched to a particular criteria of scientificity does not imply that other social organisations lack a scientific basis for their economic activities.

If sustainable utilisation is the objective that guides the criteria of choice for a development strategy, a resource prudent technological path (T.) is rationally chosen. If maximization of the growth of man-made processes and increasing the productivity of labour is the objective, then a more resource-intensive path (T2) which is the integration of a large number of smaller technologies (t1) and in which increased resource and energy inputs allows the increase in labour productivity, is rationally chosen. In this process a large amount of secondary resources (R2-R6) are additionally required.

Traditional societies in all their diversity have, in general, shared a common set of characteristics. They have used natural resources prudently to satisfy minimum needs sustainably over centuries. Such resource use was based on.

Resource flow in resource prudent t1 and resource.-intensive t2 technology chains

1. A knowledge system with an ecological understanding of nature.
2. A technological system for processing resources to satisfy human needs with minimum resource waste.

3. Rationality criteria for demarcating vital and non-vital needs and between resource destructive and resource enhancing technologies.

Traditional world views and practices deterred over-exploitation of natural resources at all levels. As they were based on ecological perceptions of nature and guided by restraints in resource use, they used technologies which prevented ecological disruption.

Modernisation of traditional societies in its present form has, by and large, been taken as synonymous with the substitution of indigenous science and technology systems by the modern western system. In this manner the resource-intensive western pattern of resource use is thrust on non-western societies through modernisation.

Modern western scientific knowledge, however, differs from indigenous knowledge systems in three important ways:

1. Modern western scientific knowledge is reductionist and fragmented.
2. Modern western technological systems are based on reductionist science and are generally more resource-intensive.
3. There are no criteria of rationality or technology choices to evaluate modern science and technology on the basis of resource use efficiency or need satisfaction capability.

These characteristics of modern western science and technology systems breaks the chain, beginning with natural resources and ending in the satisfaction of human needs and demands, into small fragments of individually identifiable economic activities. This provides justification for the resource intensity of the dominant paradigm of economic development and technological change, and thus leads to ecological instabilities.

Ecological crises are thus inevitable products of economic activities which are propelled towards longer and more complex and resource-intensive technological chains (T2) for the satisfaction of older needs (N.). Only individual segments(l) of the whole technological chain are examined from the narrow criteria of labour productivity. The situation is best exemplified in the case of food production. While indigenous and traditional food production practices used about half a calorie of energy to produce 1 calorie of food, the present mechanised and chemical farming techniques use 10 calories of energy to produce 1 calorie of food. These characteristics of contemporary scientific industrial development are the primary causes for the contemporary ecological crises. The combination of ecologically disruptive scientific and technological modes, and the absence of rationality criteria for evaluating scientific and technological systems in terms of resource use efficiency, has created conditions where society is increasingly propelled towards ecological instability and has no rational and organised response to arrest and curtail these destructive tendencies.

PRINCIPLE OF ECOLOGY AND ECOSYSTEM

The first principle of ecology is that each living organism has an ongoing and continual relationship with every other element that makes up its environment. An ecosystem can be defined as any situation where there is interaction between organisms and their environment.

The ecosystem is of two entities, the entirety of life, the biocoenosis, and the medium that life exists in, the biotope. Within the ecosystem, species are connected by food chains or food webs. Energy from the sun, captured by primary producers via photosynthesis, flows upward through the chain to primary consumers (herbivores), and then to secondary and tertiary consumers (carnivores and omnivores), before ultimately being lost to the system as waste heat. In the process, matter is incorporated into living organisms, which return their nutrients to the system via decomposition, forming biogeochemical cycles such as the carbon and nitrogen cycles. The concept of an ecosystem can apply to units of variable size, such as a pond, a field, or a piece of dead wood. An ecosystem within another ecosystem is called a micro ecosystem. For example, an ecosystem can be a stone and all the life under it. A meso ecosystem could be a forest, and a macro ecosystem a whole eco region, with its drainage basin. The main questions when studying an ecosystem are:

- Whether the colonization of a barren area could be carried out
- Investigation the ecosystem's dynamics and changes
- The methods of which an ecosystem interacts at local, regional and global scale
- Whether the current state is stable
- Investigating the value of an ecosystem and the ways and means that interaction of ecological systems provides benefits to humans, especially in the provision of healthy water.

Ecosystems are often classified by reference to the biotopes concerned. The following ecosystems may be defined:

- As continental ecosystems, such as forest ecosystems, meadow ecosystems such as steppes or savannas, or agro-ecosystems
- As ecosystems of inland waters, such as lentic ecosystems such as lakes or ponds; or lotic ecosystems such as rivers
- As oceanic ecosystems.

Another classification can be done by reference to its communities, such as in the case of an human ecosystem.

THE ECONOMY OF NATURAL ECOLOGICAL PROCESSES

The terms ecology and economy are rooted in the same Greek word 'oikos' or household. Yet in the context of market-oriented development they have been rendered contradictory: 'Ecological destruction is an obvious cost for economic development'-a statement which is often repeated to ecology movements. Natural resources are produced and reproduced through a complex network of ecological processes. Production is an integral part of this economy of natural ecological processes but the concepts of production and productivity in the context of development economics have been exclusively identified with the industrial production system for the market economy. Organic productivity in forestry or agriculture has also been viewed narrowly through the production of marketable products of the total productive process.

This has resulted in vast areas of resource productivity, like the production of humus by forests, or regeneration of water resources, natural evolution of genetic products, erosional production of soil fertility from parent rocks, remaining beyond the scope of economics. Many of these productive processes are dependent on a number of ecological processes. These processes are not known fully even within the natural science disciplines and economists have to make tremendous efforts to internalize them. Paradoxically, through the resource ignorant intervention of economic development at its present scale, the whole natural resource system of our planet is under threat of a serious loss of productivity in the economy of natural processes.

At present ecology movements are the sole voice to stress the economic value of these natural processes. The market-oriented development process can destroy the economy of natural processes by over exploitation of resources or by the destruction of ecological processes that are not comprehended by economic development. And these impacts are not necessarily manifested within the period of the development projects. The positive contribution of economic growth from such development may prove totally inadequate to balance the invisible or delayed negative externalities stemming from damage to the economy of natural ecological processes. In the larger context, economic growth can thus, itself become the source of underdevelopment. The ecological destruction associated with uncontrolled exploitation of natural resources for commercial gains is a symptom of the conflict between the ways of generating material wealth in the economies of-market and the natural processes. In the words of Commoner: 'Human beings have broken out of the circle of life driven not by biological needs, but the social organisation which they have devised to 'conquer' nature: means of gaining wealth which conflict with those which govern nature."

7

Rock Geology

In geology, a rock is a naturally occurring solid aggregate of one or more minerals or mineraloids. For example, the common rock granite is a combination of the quartz, feldspar and biotite minerals. The Earth's outer solid layer, the lithosphere, is made of rock.

Fig. Balanced Rock ***stands in Garden of the Gods park in Colorado Springs, CO.***

Rocks have been used by mankind throughout history. From the Stone Age rocks have been used for tools. The minerals and metals we find in rocks have been essential to human civilization.

Three major groups of rocks are defined: igneous, sedimentary, and metamorphic. The scientific study of rocks is called petrology, which is an essential component of geology.

BASIC ROCK MECHANICS

This module is intended to provide a basic understanding of the engineering principles used in the analysis of rock materials when these materials are intended to be used for an engineering purpose. These principles form the foundation of the science of rock mechanics and have been widely adopted and used. At the introductory level provided here, the applicability of these principles should be easily grasped by each student. These notes are, in part, from my course Geology for Engineers, GeoE 221, a first-level course in geology. The course is presented as a systematic study of Earth and how natural geologic processes are applied in engineering practice. This course is a freshman-level Geological Engineering course but is also taken by students of all grades from other disciplines such as Civil and Environmental Engineering and Mechanical Engineering.

Rock Basics

A. Engineering Uses: Rock is used for engineering purposes in 2 primary ways:

1. *as a building material:* items such as cut stones, beams, support columns, decorative panels, etc. Each student can envision examples where rock has been used in one of these ways.

2. *as a foundation:* For example, on Manhattan Island, the skyscapers are founded on granite. In Central Park, just a short distance to the south, there are no buildings over a few stories. Why? The bedrock under the Park consists of marine shale and metamorphic rocks that will not support the weight of a skyscraper.

Thus, knowing and understanding basic rock properties will allow structures to be founded correctly so the required support will be there.

B. Rock Measurements: the physical characteristics of a rock mass are a fundamental geologic property and are extremely important to engineers. Analytical data on theses characteristics are generally derived in 2 ways:

1. *laboratory measures:* are generally referred to as 'rock properties' and are acquired using small samples taken from the field site and analyzed in a laboratory setting.

2. *field-scale measures:* most often referred to as 'rock mass properties' and are descriptions of the bulk strength properties of the rock mass. The nature of these properties are governed primarily by 'discontinuities', or planes of weakness, that are present in the rock mass. Examples of discontinuities are fractures, bedding planes, faults, etc. The measured distance between fractures, bedding planes, and other structural features are also important when collecting field-scale data.

C. Definitions: From an engineering standpoint, there is a difference between a rock and a stone. There is also a difference between soil and dirt. 'Rock' is used to denote the mass of material in-situ, it is a part of the bedrock and has not been moved or disturbed. 'Stone' is used to denote rock material that has been removed from its bedrock location. In the same way, 'soil' refers to that material naturally in place and 'dirt' is most often used to define soil that has been removed and is, or has, been transported.

Mechanical Properties of Rock Material

Compressive Strength: Compressive strength is the capacity of a material to withstand axially directed compressive forces. The most common measure of compressive strength is the uniaxial compressive strength or unconfined compressive strength. Usually compressive strength of rock is defined by the ultimate stress. It is one of the most important mechanical properties of rock material, used in design, analysis and modelling.

Stage I – The rock is initially stressed, pre-existing microcracks or pore orientated at large angles to the applied stress is closing, in addition to deformation. This causes an initial non-linearity of the axial stress-strain curve. This initial non-linearity is more obvious in weaker and more porous rocks,

Stage II – The rock basically has a linearly elastic behaviour with linear stress-strain curves, both axially and laterally. The Poisson's ratio, particularly in stiffer unconfined rocks, tends to be low. The rock is primarily undergoing elastic deformation with minimum cracking inside the material. Micro-cracks are likely initiated at the later portion of this stage, of about 35-40% peak strength. At this stage, the stress-strain is largely recoverable, as the there is little permanent damage of the micro-structure of the rock material.

Stage III – The rock behaves near-linear elastic. The axial stress-strain curve is nearlinear and is nearly recoverable. There is a slight increase in lateral strain due to dilation. Microcrack propagation occurs in a stable manner during this stage and that microcracking events occur independently of each other and are distributed throughout the specimen. The upper boundary of the stage is the point of maximum compaction and zero volume change and occurs at about 80% peak strength.

Stage IV – The rock is undergone a rapid acceleration of microcracking events and volume increase. The spreading of microcracks is no longer independent and clusters of cracks in the zones of highest stress tend to coalesce and start to form tensile fractures or shear planes - depending on the strength of the rock.

Stage V – The rock has passed peak stress, but is still intact, even though the internal structure is highly disrupt. In this stage the crack arrays fork and coalesce into macrocracks or fractures. The specimen is undergone strain softening (failure) deformation, i.e., at peak stress the test specimen starts to become weaker with increasing strain. Thus further strain will be concentrated on weaker elements of the rock which have already been subjected to strain. This in turn will lead to zones of concentrated strain or shear planes.

Stage VI – The rock has essentially parted to form a series of blocks rather than an intact structure. These blocks slide across each other and the predominant deformation mechanism is friction between the sliding blocks. Secondary fractures may occur due to differential shearing. The axial stress or force acting on the specimen tends to fall to a constant residual strength value, equivalent to the frictional resistance of the sliding blocks. In underground excavation, we often are interested in the rock at depth. The rock is covered by overburden materials, and is subjected to lateral stresses. Compressive strength with lateral pressures is higher than that without. The compressive strength with lateral pressures is called triaxial compressive strength. The true triaxial compression means the 3 different principal stresses in three directions. A true triaxial compression testing machine is rather difficult to operate. In most tests, two lateral stresses are made equal, i.e., $\sigma 2 = \sigma 3$. In the test, only a confining pressure is required.

Young's Modulus and Poisson's Ratio: Young's Modulus is modulus of elasticity measuring of the stiffness of a rock material. It is defined as the ratio, for small strains, of the rate of change of stress with strain. This can be experimentally determined from the slope of a stress-strain curve obtained during compressional or tensile tests conducted on a rock sample. Similar to strength, Young's Modulus of rock materials varies widely with rock type. For extremely hard and strong rocks, Young's Modulus can be as high as 100 GPa. There is some correlation between compressive strength and Young's Modulus, and discussion is given in a later section. Poisson's ratio measures the ratio of lateral strain to axial strain, at linearly-elastic region. For most rocks, the Poisson's ratio is between 0.15 and 0.4. As seen from early section, at later stage of loading beyond linearly elastic region, lateral strain increase fast than the axial strain and hence lead to a higher ratio.

Stress-Strain at and after Peak: Strain at failure is the strain measured at ultimate stress. Rocks generally fail at a small strain, typically around 0.2 to 0.4% under uniaxial compression. Brittle rocks, typically crystalline rocks, have low strain at failure, while soft rock, such as shale and mudstone, could have relatively high strain at failure. Strain at failure sometimes is used as a measure of brittleness of the rock. Strain at failure increases with increasing confining pressure under triaxial compression conditions. Rocks can have brittle or ductile behaviour after peak. Most rocks, including all crystalline igneous, metamorphic and sedimentary rocks, behave brittle under uniaxial compression. A few soft rocks, mainly of sedimentary origin, behave ductile.

Tensile Strength: Tensile strength of rock material is normally defined by the ultimate strength in tension, i.e., maximum tensile stress the rock material can withstand. Rock material generally has a low tensile strength. The low tensile strength is due to the existence of microcracks in the rock. The existence of microcracks may also be the cause of rock failing suddenly in tension with a small strain. Tensile strength of rock materials can be obtained from several types of tensile tests: direct tensile test, Brazilian test and flexure test. Direct test is not commonly performed due to the difficulty in sample preparation. The most common tensile strength determination is by the Brazilian tests.

Shear Strength: Shear strength is used to describe the strength of rock materials, to resist deformation due to shear stress. Rock resists shear stress by two internal mechanisms, cohesion and internal friction. Cohesion is a measure of internal bonding of the rock material. Internal friction is caused by contact between particles, and is defined by the internal friction angle, ϕ. Different rocks have different cohesions and different friction angles. Shear strength of rock material ca be determined by direct shear test and by triaxial compression tests. In practice, the later methods is widely used and accepted. With a series of triaxial tests conducted at different confining pressures, peak stresses ($\sigma 1$) are obtained at various lateral stresses ($\sigma 3$).

Tensile and shear strengths are important as rock fails mostly in tension and in shearing, even the loading may appears to be compression. Rocks generally have high compressive strength so failure in pure compression is not common.

PROPERTIES OF ROCKS AND SOILS

Rocks are important. They are the things that the Earth is made of so students need to learn about them. Geologists classify rocks into groups depending on how they are formed and what their major components are. Many rocks can be thought of in the same terms used when someone is baking cookies. You have the batter and maybe put pieces of chocolate or nuts in the batter to make cookies. Then they are baked in the oven.

Rocks are made in similar fashion. They are composed of various minerals which are in turn composed of chemical elements. These minerals form the rocks which generally are produced under tremendous heat and pressure conditions deep underground.

Rocks can be classified according to their properties. One suggestion is given in Key to Rocks. This lists properties of a variety of rocks and sorts the rocks according to their properties.

The Rock Cycle

Many natural phenomena occur in cycles. The rock cycle explains the relationship between the three major groupings of rocks.

- Igneous rocks are formed deep within the Earth. They come up to the surface underground (in the form called magma) or break out of the ground (in the form called lava).
- Sedimentary rocks are formed in layers as sediments, usually on the bottom of the ocean. Sometimes sedimentary rocks begin as layers of

dirt or sand on the land. They get buried under thousands and thousands of feet of other rocks and turn to stone.

- Metamorphic rocks begin as either igneous rocks or sedimentary rocks. Through the processes associated with plate tectonics, these rocks are taken deep into the ground by subduction. Once deep underground, they are heated and remelted. They later cool and become metamorphic rocks.

Formation of rocks takes a long, long time. Many of the rocks in Northwest New Jersey are well over 200 million years old. Many others go back much longer than that—half a billion years or more. Scientists use theGeologic Time Scale to indicate the ages of rocks.

DEFORMATION OF ROCK

Within the Earth rocks are continually being subjected to forces that tend to bend them, twist them, or fracture them. When rocks bend, twist or fracture we say that they deform (change shape or size). The forces that cause deformation of rock are referred to as stresses (Force/unit area). So, to understand rock deformation we must first explore these forces or stresses.

Stress and Strain

Stress is a force applied over an area. One type of stress that we are all used to is a uniform stress, called pressure. A uniform stress is a stress wherein the forces act equally from all directions. In the Earth the pressure due to the weight of overlying rocks is a uniform stress, and is sometimes referred to as confining stress.

If stress is not equal from all directions then we say that the stress is a differential stress. Three kinds of differential stress occur.

1. *Tensional stress (or extensional stress),* which stretches rock;
2. *Compressional stress,* which squeezes rock; and
3. *Shear stress,* which result in slippage and translation.

When rocks deform they are said to *strain*. A strain is a change in size, shape, or volume of a material.

Stages of Deformation

When a rock is subjected to increasing stress it passes through 3 successive stages of deformation.

- *Elastic Deformation* — wherein the strain is reversible.
- *Ductile Deformation* — wherein the strain is irreversible.
- *Fracture* - irreversible strain wherein the material breaks.

We can divide materials into two classes that depend on their relative behavior under stress.

- Brittle materials have a small or large region of elastic behavior but only a small region of ductile behavior before they fracture.
- Ductile materials have a small region of elastic behavior and a large region of ductile behavior before they fracture.

How a material behaves will depend on several factors. Among them are:

- Temperature - At high temperature molecules and their bonds can stretch and move, thus materials will behave in more ductile manner. At low Temperature, materials are brittle.
- Confining Pressure - At high confining pressure materials are less likely to fracture because the pressure of the surroundings tends to hinder the formation of fractures. At low confining stress, material will be brittle and tend to fracture sooner.
- Strain rate — At high strain rates material tends to fracture. At low strain rates more time is available for individual atoms to move and therefore ductile behavior is favored.
- Composition — Some minerals, like quartz, olivine, and feldspars are very brittle. Others, like clay minerals, micas, and calcite are more ductile This is due to the chemical bond types that hold them together. Thus, the mineralogical composition of the rock will be a factor in determining the deformational behavior of the rock. Another aspect is presence or absence of water. Water appears to weaken the chemical bonds and forms films around mineral grains along which slippage can take place. Thus wet rock tends to behave in ductile manner, while dry rocks tend to behave in brittle manner.

Brittle-Ductile Properties of the Lithosphere

We all know that rocks near the surface of the Earth behave in a brittle manner. Crustal rocks are composed of minerals like quartz and feldspar which have high strength, particularly at low pressure and temperature. As we go deeper in the Earth the strength of these rocks initially increases. At a depth of about 15 km we reach a point called the brittle-ductile transition zone.

Below this point rock strength decreases because fractures become closed and the temperature is higher, making the rocks behave in a ductile manner. At the base of the crust the rock type changes to peridotite which is rich in olivine. Olivine is stronger than the minerals that make up most crustal rocks, so the upper part of the mantle is again strong. But, just as in the crust, increasing temperature eventually predominates and at a depth of about 40 km the brittle-ductile transition zone in the mantle occurs. Below this point rocks behave in an increasingly ductile manner.

DEFORMATION IN PROGRESS

Only in a few cases does deformation of rocks occur at a rate that is observable on human time scales. Abrupt deformation along faults, usually associated with earthquakes caused by the fracture of rocks occurs on a time scale of minutes or seconds. Gradual deformation along faults or in areas of uplift or subsidence can be measured over periods of months to years with sensitive measuring instruments.

Evidence of Former Deformation

Evidence of deformation that has occurred in the past is very evident in crustal rocks. For example, sedimentary strata and lava flows generally follow the law of original horizontality. Thus, when we see such strata inclined instead of horizontal, evidence of an episode of deformation. In order to uniquely define the orientation of a planar feature we first need to define two terms- strike and dip.

For an inclined plane the *strike* is the compass direction of any horizontal line on the plane. The *dip* is the angle between a horizontal plane and the inclined plane, measured perpendicular to the direction of strike.

In recording strike and dip measurements on a geologic map, a symbol is used that has a long line oriented parallel to the compass direction of the strike. A short tick mark is placed in the center of the line on the side to which the inclined plane dips, and the angle of dip is recorded next to the strike and dip symbol as shown above.

ACTIVITY: PROPERTIES OF ROCKS

Rocks have a variety of properties that can be studied in the classroom. Among others, these include

- color: some rocks are dark, others are light
- crystals: some rocks have crystals and others do not. If they do have crystals
    - o arrangement: are the crystals random or in layers
    - o similarity: are the crystals uniform or different
- particles: some rocks contain sand or gravel particles
- holes: many volcanic rocks have holes in them
- layers: sedimentary rocks have layers as do some metamorphic rocks
- reflectivity: some rocks reflect light like a mirror
- luster: some rocks are shiny, others are dull
- hardness: the Mohs hardness scale is used
- magnetic properties: some rocks attract a magnet, others do not

Geologists also use several other properties in studying rocks. These include the density and specific gravity, reactivity with acid and the color of the streak that the rock leaves on a piece of unglazed tile.

Materials

- rock samples. This could include a variety of rocks found in various parts of New Jersey. It is possible to obtain kits which contain a rocks with a variety of hardness levels for testing.
- large nail
- glass plate. A glass company can make some of these for you. As you will be scratching them with rocks, you do not want to use the school's windows for this. Any glass will do but it has to be something that you do not mind the students scratching all up.
- magnets
- pennies
- magnifiers
- Science Journals

ACTIVITY: PROPERTIES OF SOILS

Soil is a mixture. It contains organic material such as decayed leaves, often called humus. It also contains rock particles produced by weathering of rocks and rock formations. Different types of soils are recognized based on their contents.

- Sandy soil contains a large proportion of sand in addition to other materials.
- Loamy soil or loam contains a large proportion of organic material or humus.
- Clayey soil or clay contains a large proportion of clay made of very small rock particles.

Many times the properties of the soil are determined by the properties of the rocks below them.

Much of the really good farm land in south central Pennsylvania has soil that is directly above limestone deposits. The limestone adds lime to the soil.

Materials

- soil samples (garden soil, sandy soil, clay)
- magnifiers
- metric rulers
- toothpicks
- Science Journals

Procedure.

1. Students describe each of the three types of soils in their Science Journals. They make lists of the properties of the soils they are studying and compare and contrast them.
2. Students examine a sample of each type of soil with a magnifier. They carefully tease apart the soil particles with toothpicks to get a good close look at them. They write their observations in their Science Journals.
3. Students describe where each type of soil is found and what it may be used for.

PHYSICAL PROPERTIES OF ROCKS

Permeability of Rocks

Stereographic projection of the electrical conductivity of undeformed and deformed Carrara marble

An experimental study of water permeability of mudstones has been carried out to evaluate some of the complexities of fluid flow through clay-bearing rocks (David Hawthorn, PhD graduate, with Ernie Rutter and Joe MacQuaker). We have also investigated the formation of faulting during compaction of muddy rocks, to help understand the formation of polygonal fault arrays in nature (Alex McDougall (MSc graduate) and Ernie Rutter).

We have collected oriented samples of clay-bearing fault gouge and measured its anisotropic permeability to water and inert gas, as a function of pore pressure, confining pressure, temperature and differential stress. The pore pressure oscillation techniques has been extensively employed for this. Permeability evolution has also been studied during progressive shearing of synthetic clay-bearing gouges of different clay content (Brian Crawford (now with Exxon-Mobil), Dan Faulkner (now at Liverpool) and Ernie Rutter).

Seismic Velocity Measurements

As part of our ongoing studies of lower crustal rocks, we have taken a full range of oriented cores from the Ivrea-Verbano lower crustal section of N. Italy and measured ultrasonic velocities as a function of pressure and temperature to 700 C and 500 MPa. These data have been combined with a reconstructed cross section through the region as it was when it was in the lower crust to produce a synthetic seismic reflection model of the section, for comparison with contemporary lower crustal reflection profiles (collaboration with Dave Waltham and Derek Blundell, Royal Holloway college, London). A new pressure vessel exclusively dedicated to high pressure velocity measurements is currently being used (Steve Covey-Crump, Ernie Rutter and

Chidi Chukwunweike) to characterize velocities in a range of reservoir rocks at elevated confining and pore pressure.

INTRUSIVE IGNEOUS ROCKS

Pluton - Body of magma which has solidified beneath the earth. Classified based on whether they are concordant (i.e. they are parallel to layering of host) or discordant (cross cut host). Also if they are tabular or massive (equi-dimensional football-shaped).

1. *Sill* - Tabular concordant pluton
2. *Dike* - Tabular discordant pluton
3. *Laccolith* - Massive concordant pluton
4. *Batholith* - Massive discordant pluton.

Magma Crystallization

By the end of the 19th century it was recognized that all igneous rocks formed from the crystallization of a magma. A fundamental question that followed was "why do we get so many different types of igneous rocks if we had one primordial starting material". Use the analogy of baking a cake. N.L. Bowen conducts the first systematic study of the crystallization of igneous rocks.

Publishes Bowen's Reaction Series which shows that the minerals in igneous rocks crystallize in an orderly sequence. Discontinuous Series so named because as temperature falls we change from one new mineral to another (Ex. olivine alters to pyroxene). Continuous Series in which plagioclase feldspar merely changes composition from Ca-rich at high temperature to Na-rich at low temperature. Does not involve the formation of a new mineral, just a compositional change. This does not really help us understand why we have different igneous rocks, but it does seem to show that there is some order in nature. To more closely examine this order let's look only at the plagioclase feldspars. Why? Because plagioclase occurs in most igneous rocks. So if we can understand how and why feldspars form we may have some understanding about how different rocks form.

Plot of temperature vs. composition. Upper line is liquidus. Separates *liquid* field from *liquid + crystals* field. Lower line is the solidus which separates the *liquid + crystals* field from the *solid* field. We can begin by examining the crystallization path of a liquid of composition X_0. It cools to temperature X_1 and at that point the first crystals begin to form. To determine their composition we project a horizontal line to the solidus and find they have the composition C_1 or about 85% Ca plag. As temperature continues to fall liquid composition shifts along liquidus to X_2 and solid crystals shift in composition along the solidus to C_2. At the completion of crystallization, (about 1275°C) the final solid has exactly the same composition as the starting liquid. This is an example of equilibrium crystallization.

Theoretically, fractional crystallization seems possible, but how could it occur in nature? By the process of gravitative settling, in which the early formed crystals in a magma sink to the bottom of the chamber due to their greater density and as such are shielded from reacting with the magma. Result is a series of layers of crystals of differing composition. Where can we find such a phenomenon in nature? The Palisades Sill that has occurred as the result of gravitative settling and fractional crystallization.

We could form each of these rocks as the result of fractional crystallization. The problem with fractional crystallization, however, is that it is not very efficient. Even under the best of circumstances we can form only 5% granite by fractional crystallization. Continents are 60% granite so where did all of it come from? Answer is there must be another mechanism involved. Go back to Plagioclase Phase Diagram and look at what happens if we take a solid of 50% Na plagioclase and 50% Ca plagioclase and heat it just enough to partially melt the solid. Liquid that forms is very Na-rich. Because it is a liquid it rises out of the system, eventually to crystallize higher in the crust. The solid that forms has the very same Na-rich plagioclase as the composition of the liquid. Thus if we partially melt a solid we can generate a liquid of very different composition which eventually recrystallizes as a rock of very different composition. This mechanism of forming rocks of different composition is termed Partial Melting and is thought to be the dominant mode of formation of the various different rocks.

Partial melting leads to the following:

peridotite —> basalt

basalt —> andesite

andesite —> granite (rhyolite)

Mantle of the earth thought to be peridotite. This conclusion ts based on the velocity of seismic waves and samples of peridotite found in diamond pipes. If we partially melt a peridotite (3-8%) the magma we generate has the composition of a basalt. The typical result of partial melting of mantle peridotite at a divergent plate boundary such as the Mid- Atlantic Ridge. The crust is pulled apart and a basaltic magma is produced and then rises upward and emplaces itself on the sea floor as a pillow lava. Beneath the pillow lavas are diabase dikes, gabbro and peridotite.

The situation is different for the formation of granites at subduction zones. In order to form a partial melt at realistic depths we need water. This is because water dramatically lowers the melting point of rocks. The water comes from sediments carried down the subduction zone at convergent plate boundaries. Water lowers melting point of sediments and surrounding igneous rocks, thus forming a partial melt at 30-50km.

So the following occur:

At divergent plate boundaries peridotite mantle partially melts to give basalt magma.

At convergent plate boundaries water is carried down subduction zones causing partial melting and the formation of granitic magmas.

CLASSIFICATION OF ROCK

At a granular level, rocks are composed of grains of minerals, which, in turn, are homogeneous solids formed from a chemical compound that is arranged in an orderly manner. The aggregate minerals forming the rock are held together by chemical bonds. The types and abundance of minerals in a rock are determined by the manner in which the rock was formed. Many rocks contain silica (SiO_2); a compound of silicon and oxygenthat forms 74.3% of the Earth's crust. This material forms crystals with other compounds in the rock. The proportion of silica in rocks and minerals is a major factor in determining their name and properties.

Rocks are geologically classified according to characteristics such as mineral and chemical composition,permeability, the texture of the constituent particles, and particle size. These physical properties are the end result of the processes that formed the rocks. Over the course of time, rocks can transform from one type into another, as described by the geological model called the rock cycle. These events produce three general classes of rock: igneous, sedimentary, and metamorphic.

The three classes of rocks are subdivided into many groups. However, there are no hard and fast boundaries between allied rocks. By increase or decrease in the proportions of their constituent minerals they pass by every gradation into one another, the distinctive structures also of one kind of rock may often be traced gradually merging into those of another. Hence the definitions adopted in establishing rock nomenclature merely correspond to more or less arbitrary selected points in a continuously graduated series.

Igneous

Igneous rock (derived from the Latin word *igneus* meaning of fire, from *ignis* meaning fire) forms through the cooling and solidification of magma or lava. This magma can be derived from partial melts of pre-existing rocks in either a planet's mantle or crust. Typically, the melting of rocks is caused by one or more of three processes: an increase in temperature, a decrease in pressure, or a change in composition. Igneous rocks are divided into two main categories: plutonic rock and volcanic. Plutonic or intrusive rocks result when magma cools and crystallizes slowly within the Earth's crust. A common example of this type isgranite. Volcanic or extrusive rocks result from magma reaching the surface either as lava or *fragmental ejecta,* forming minerals such as pumice or basalt. The chemical abundance and the rate of cooling of magma typically forms a sequence known as Bowen's reaction series, after the Canadian petrologist Norman L. Bowen. Most major igneous rocks are found along this scale.

About 64.7% of the Earth's crust by volume consists of igneous rocks; making it the most plentiful category. Of these, 66% are basalts and gabbros, 16% are granite, and 17% granodiorites and diorites. Only 0.6% are syenites and 0.3% peridotites and dunites. The oceanic crust is 99% basalt, which is an igneous rock of mafic composition. Granites and similar rocks, known as meta-granitoids, form much of the continental crust. Over 700 types of igneous rocks have been described, most of them having formed beneath the surface of Earth's crust. These have diverse properties, depending on their composition and how they were formed.

Sedimentary

Sedimentary rocks are formed by sedimentation of particles at or near the Earth's surface and within bodies of water. This process causes clastic sediments or organic particles (detritus) to settle and accumulate, or for minerals to chemically precipitate (evaporite) from a solution. The particulate matter then undergoes compaction and cementation during diagenesis.

Before being deposited, sediment was formed by weathering and erosion in a source area, and then transported to the place of deposition by water, wind, ice, mass movement or glaciers which are called agents of denudation. Mud rocks comprise 65% (mudstone, shale and siltstone); sandstones 20 to 25% andcarbonate rocks 10 to 15% (limestone and dolostone). About 7.9% of the crust by volume is composed of sedimentary rocks, with 82% of those being shales, while the remainder consist of limestone (6%), sandstone and arkoses (12%).

Metamorphic

Metamorphic rocks are formed by subjecting any rock type—sedimentary rock, igneous rock or another older metamorphic rock—to different temperature and pressure conditions than those in which the original rock was formed. This process is called metamorphism; meaning to "change in form". The result is a profound change in physical properties and chemistry of the stone. The original rock, known as the protolith, transforms into other mineral types or else into other forms of the same minerals, such as by recrystallization. The temperatures and pressures required for this process are always higher than those found at the Earth's surface: temperatures greater than 150 to 200 °C and pressures of 1500 bars. Metamorphic rocks compose 27.4% of the crust by volume.

The three major classes of metamorphic rock are based upon the formation mechanism. An intrusion of magma that heats the surrounding rock causes contact metamorphism—a temperature-dominated transformation. Pressure metamorphism occurs when sediments are buried deep under the ground; pressure is dominant and temperature plays a smaller role. This is termed burial metamorphism, and it can result in rocks such as jade. Where both heat and pressure play a role, the mechanism is termed regional metamorphism. This is typically found in mountain-building regions.

Depending on the structure, metamorphic rocks are divided into two general categories. Those that possess a texture are referred to as foliated; the remainder are termed non-foliated. The name of the rock is then determined based on the types of minerals present. Schists are foliated rocks that are primarily composed oflamellar minerals such as micas. A gneiss has visible bands of differing lightness, with a common example being the granite gneiss. Other varieties of foliated rock include slates, phyllites, and mylonite. Familiar examples of non-foliated metamorphic rocks include marble, soapstone, and serpentine. This branch contains quartzite—a metamorphosed form ofsandstone—and hornfels.

CLASSIFICATION OF METAMORPHIC ROCKS

The classification and naming of metamorphic rocks is a difficult undertaking. Unlike sedimentary and igneous rocks, that have very straightforward and easy to follow classification methods the metamorphic rocks are named based on their texture and mineralogy. And because of the wide range of different temperatures and pressure combinations under which metamorphic rocks form there are literally thousands of different names given to metamorphic rocks. So, should we toss in our rock hammers, throw our hands in the air and give up? While this might make us feel better, it won't help us identify metamorphic rocks. But not to worry, despite these seemingly insurmountable odds, geologists have come up with a way to name and classify metamorphic rocks that is both simple and easy to use, once you master the basics.

Most metamorphic rocks are named based on two things, their texture and structural features, and their mineralogy. Both methods are often used together to give each rock a distinct name that allows other geologists to know exactly what type of rock is being described. Textural and structural features are often determined by the changes in pressure and temperature. We briefly covered these aspects in the previous article, but let us cover them again to make sure the concept is understood. Texture is classified by how the parent rock is changed when stress is applied to it. When the stress is applied evenly, resulting in confining pressure, the parent rock often undergoes recrystallization but often no structural changes are seen. These types of metamorphic rocks are referred to as non-foliated rocks. When the stresses are unequal, called differential stress, the parent rock will undergo recrystallization and re-alignment of the minerals into an orientation parallel to the direction of stress. These types of metamorphic rocks are said to have a planar structure and are referred to as foliated. Knowing and identifying the differences between non-foliated and foliated rocks is the first step in classifying metamorphic rocks.

Among the foliated textures, metamorphic rocks are further classified based on how the minerals within the parent rock are affected by the changes in temperature and pressure. In general terms, increases in temperature and pressure result in different, more complex textures, and different metamorphic rocks. The diagram at the right demonstrates this for a fine-grained mudrock. At the lowest amount of metamorphism, the minerals are aligned along a preferred direction in response to the applied stresses. The minerals within the parent rock often are not recrystallized and remain microscopic. The resulting rock can be easily split along cleavage planes and is called a slate. As metamorphism continues, the minerals remain in the same orientation, and certain minerals like micas and chlorites increase in grain size. This results in a fine-grained rock that is similar to slate but often develops a silky sheen on cleavage surfaces and is called a phyllite. As metamorphism continues the minerals become large enough that we can see the crystals without the aid of a hand lens. Some platy minerals, like mica, develop a strong preferred orientation and is now called a schist. Schists often develop porphyroblasts, crystals of one or two minerals present in the rock that are distinctly larger than the average crystals in the rock. As more metamorphism occurs the minerals in the rock begin to separate out into medium to coarse bands of different mineralogy and texture, often with separation of the lighter coloured minerals from the darker coloured ones. The rock is now called a gneiss.

This generalized classification, from slate to phyllite to schist to gneiss, if often enough for quick field identification, but closer examination of the mineralogy will give you a more detailed and descriptive name. Often the major mineral component(s) of the rock will be added to the rock name. If the major mineral components of a schist were mica and garnet, then the rock would be called a mica garnet schist. A gneiss with large amounts of biotite would be called a biotite gneiss. Sometimes, if the metamorphic rock has a similar mineral composition of an igneous rock, then the rock is given an appropriate descriptor like granite gneiss for a gneiss with a mineral composition similar to that of a granite.

Non-foliated metamorphic rocks are usually named exclusively on the basis of mineral composition. A marble is a granoblastic metamorphic rock composed exclusively of calcite or dolomite. (Granoblastic is a term to describe a metamorphic rock with crystals that are all nearly the same size.) Similarly a quartzite is a granoblastic metamorphic rock composed of quartz. A few of the more common names applied to different metamorphic rocks:

Amphibolite: A medium-to coarse-grained metamorphic rock composed of an amphibole (like hornblende) and plagioclase.

Eclogite: A medium-grained metamorphic rock composed of pyroxene and pyrope-rich garnet. Eclogites share the same mineral composition as basalts.

Granulite: An even-grained metamorphic rock composed of granoblastic minerals like quartz and feldspar.

Hornfels: A fine-grained, thermally metamorphosed rock (no pressure changes were applied) composed of minerals with equidimensional grains in a random orientation.

Serpentinite: A rock composed almost exclusively of minerals from the serpentine group of minerals.

Skarn: An impure marble that contains calcium-silicate minerals such as garnet and epidote.

This list of metamorphic rocks is by no means complete. As with other metamorphic rocks the predominant one or two minerals in the rock can be added to the name as a descriptor, such as garnet hornfels.

The naming and classification of metamorphic rocks is dependent on the textural and structural features of the rock and the rocks mineralogy. The first step in naming a metamorphic rock is to always identify whether the rock is foliated or non-foliated. Once this determination has been made a close examination of the rock will reveal the dominant one or two minerals within the rock. These two steps can then be combined to generate the name for the metamorphic rock.

HUMAN USE

The use of rocks has had a huge impact on the cultural and technological development of the human race. Rocks have been used by humans and other hominids for more than 2 million years. Lithic technology marks some of the oldest and continuously used technologies. The mining of rocks for their metal ore content has been one of the most important factors of human advancement, which has progressed at different rates in different places in part because of the kind of metals available from the rocks of a region.

Mining

Mining is the extraction of valuable minerals or other geological materials from the earth, from an ore body,vein or (coal) seam. This term also includes the removal of soil. Materials recovered by mining include base metals, precious metals, iron, uranium, coal, diamonds, limestone, oil shale, rock salt and potash. Mining is required to obtain any material that cannot be grown through agricultural processes, or created artificially in alaboratory or factory. Mining in a wider sense comprises extraction of any non-renewable resource (e.g.,petroleum, natural gas, or even water).

Mining of stone and metal has been done since pre-historic times. Modern mining processes involveprospecting for ore bodies, analysis of the profit potential of a proposed mine, extraction of the desired materials and finally reclamation of the land to prepare it for other uses once the mine is closed.

The nature of mining processes creates a potential negative impact on the environment both during the mining operations and for years after the mine is closed. This impact has led to most of the world's nations adopting regulations to moderate the negative effects of mining operations. Safety has long been a concern as well, though modern practices have improved safety in mines significantly.

ALKALINE/SUBALKALINE ROCKS

One last general classification scheme divides rocks that alkaline from those that are subalkaline. Note that this criteria is based solely on an alkali vs. silica. Alkaline rocks should not be confused with peralkaline rocks. While most peralkaline rocks are also alkaline, alkaline rocks are not necessarily peralkaline. On the other hand, very alkaline rocks, that is those that plot well above the dividing line are also usually silica undersaturated.

Examples of Questions on this material that could be on an Exam

1. Define the following terms: (a) Mode, (b) Norm, (c) silica saturation (d) peralkaline, (e) peraluminous, (f) metaluminous, (g) acid igneous rock.
2. Which of the following minerals, if found in a rock, would indicate that the rock is undersaturated with respect to silica? (choose all that apply)(a) nepheline, (b) leucite, (c) plagioclase, (d) quartz, (e) muscovite, (f) sodalite, (g) anorthite, (h) aegerine
3. Which of the following minerals, if found in a rock, would indicate that the rock is peraluminous? (choose all that apply) (a) nepheline,

(b) leucite, (c) plagioclase, (d) quartz, (e) muscovite, (f) corrundum, (g) kyanite, (h) reibeckite

4. Which of the following minerals, if found in a rock, would indicate that the rock is peralkaline? (choose all that apply) (a) nepheline, (b) leucite, (c) plagioclase, (d) quartz, (e) muscovite, (f) corrundum, (g) aegerine, (h) reibeckite, (i) aenigmatite.

PRINCIPLES OF ROCK DEFORMATION

Rocks deform in response to differential stress.The resulting structure depends on the stress orientation. At high temperatures, ductile flow of rocks occurs. At low temperatures, brittle fractures form.

The folds and faults exposed in canyon walls and mountain ranges show that crustal rocks can be deformed on large scales and in dramatic ways. But why are some rocks warped into great folds and others only fractured or faulted? Why are some only gently folded, whereas others are complexly folded and faulted? In short, what factors control the type of deformation that rocks experience? To understand this, you need to understand the forces on rocks. Force applied to an area is stress. Stress is the same thing as pressure and is a measure of the intensity of the force or of how concentrated the force is. Everyday experience tells us that solids will bend or break if too much stress is placed on them; that is, they deform if the stress exceeds their strength (their natural resistance to deformation). Rocks behave in the same way and deform in response to the forces applied to them.

All of Earth's rocks are under some type of stress, but in many situations the stress is equal in all directions and the rocks are not deformed. In many tectonic settings, however, the magnitude of stress is not the same in all directions and rocks experience differential stress. As a result, the rocks yield to the unequal stress and deform by changing shape or position. Geologists call the change in shape strain. In other words, differential stress causes strain.

Although strain proceeds by several complex phenomena, two end-member styles of deformation are recognized, with all gradations between them being possible. Under some conditions, rock bodies change shape by breaking to form continuous fractures and they lose cohesion; this is brittle deformation. We commonly see such behavior of solids in our daily experience: Chairs, baseball bats, pencils, and wooden beams break if too much force is applied to them. On the other hand, ductile deformation occurs when a rock body deforms permanently without fracturing or losing cohesion.The most obvious type of ductile deformation is the viscous flow of fluids, such as molten magma, but solids can also deform ductilely. At first it may seem strange that solids can flow. But bending of metal provides a familiar example.

Consider sheet metal in a car fender. A minor collision commonly makes a dent, not a fracture in the fender. Likewise, rocks can flow in a solid state, under the right conditions. This type of solid state flow is usually called plastic flow and is accomplished by slow internal creep, gliding on imperfections in crystals, and recrystallization.

Depending on the temperature or pressure of the surroundings and the rate at which stress is applied, most types of rocks can deform by brittle fracture or ductile flow. Low pressures, low temperatures, and rapid deformation rates favour brittle deformation. As a result, brittle structures are most common in the shallow crust. We use the term shear to describe slippage of one block past another on a fracture. High confining pressures, high temperatures, and low rates of deformation all favour ductile behavior. Ductile deformation is more common in the mantle and deeper parts of the crust. Another example of this difference can be seen in the behavior of glass. When a glass rod is cold, it is strong and brittle fractures form when enough stress is applied. When the same glass rod is hot but not molten, it is weak and easy to bend. To visualize the role of the rate of deformation, consider taffy (or Silly Putty™) as an example. If warm taffy is pulled slowly and steadily, it is ductile and stretches continuously, without breaking, to form long, thin strands. On the other hand, if it is stretched rapidly, it may break and form brittle fractures.

It should be clear which types of rock structures form by ductile deformation and which form by brittle behavior. The flow of rocks in a solid state to form folds in metamorphic rocks is a good example of ductile behavior. The formation of fractures, joints, and faults are common expressions of deformation in the brittle regime. We can also better understand rock structures if we consider the orientations of the stresses acting on a body. For example, tension occurs where the stresses point away from one another and tend to pull the rock body apart. In contrast, compression tends to press a body of rocks together. Three distinct types of deformation occur because of differential stresses caused by tectonic processes. To visualize this, imagine how two adjacent blocks can interact. They can move away from one another (extension), move toward one another (contraction), or slip horizontally past one another (lateral-slip). Obviously, the orientation of the stresses acting on the blocks determines which of the three cases is dominant. In the simplest case, the stress orientations are directly related to plate tectonic settings. Extension is caused when the differential stresses point away from one another. This type of deformation results in lengthening and is common at divergent boundaries. In brittle rocks it is expressed by fracturing and faulting, and in ductile rocks by stretching and thinning. Contraction is caused by horizontal compression when the differential stresses are directed toward one another. Contraction is common at convergent boundaries and causes shortening and thickening of rock bodies, expressed as faults in brittle rocks and folds in ductile rocks. Lateral-slip is the kind of shear that occurs when rocks slide horizontally past one another along nearly vertical fractures and dominates at transform plate boundaries.

ROCK MECHANICAL PROPERTIES

The determination of a reservoir's mechanical properties is critical to reducing drilling risk and maximizing well and reservoir productivity. Estimates of rock mechanical properties are central to the following :

- Drilling programs
- Well placement
- Well-completion design

Acoustic logging can provide information helpful to determining the mechanical properties of reservoir rock.

MECHANICAL PROPERTIES OF ROCK

Mechanical properties include:

- Elastic properties (Young's modulus, shear modulus, bulk modulus, and Poisson's ratio)
- Inelastic properties (fracture gradient and formation strength)

Elasticity is the property of matter that causes it to resist deformation in volume or shape. Hooke's law describes the behavior of elastic materials and states that for small deformations, the resulting strain is proportional to the applied stress.

- Stress is the force applied per unit area
- Strain is the fractional distortion that results because of the acting force
- The modulus of elasticity is the ratio of stress to strain

Depending on the mode of the acting geological force and type of geological media the force is acting upon, three types ofdeformation can result as well as three elastic moduli that correspond to each type of deformation.

- Young's modulus, *E*, is the ratio of uniaxial compressive (tensile) stress to the resultant strain
- Bulk modulus, *K*, is the change in volume under hydrostatic pressure (i.e., the ratio of stress to strain) (*K* is the reciprocal of compressibility.)
- Shear modulus, *m*, is the ratio of shearing (torsional) stress to shearing strain.
- An additional parameter, Poisson's ratio, *ó*, is a measure of the geometric change of shape under uniaxial stress.

These four elastic parameters are interrelated such that any one can be expressed in terms of two others and can also be expressed in terms of acoustic-wave velocity and density.

Computing Mechanical Rock Properties

The data needed to compute mechanical rock properties are:

- Compressional and shear velocities (slowness)
- Density

Shear and compressional velocities are a function of:

- Bulk modulus
- Shear modulus
- Density of the formation being measured

The V_p/V_s ratio, combined with formation density, ñ, is used to calculate:

- Poisson's ratio
- Young's modulus
- Bulk modulus
- Shear modulus

Whenever possible, log-derived, dynamic rock properties should be calibrated to core-derived static (laboratory) properties, because the static measurements more accurately represent the in-situ reservoir mechanical properties. Rock mechanical properties can be determined using either of the following:

- Conventional empirical charts
- Computer programs

The elastic moduli and Poisson's ratio are used in a variety of applications. These applications include:

- Predictions of formation strength
- Well stimulation (fracture pressure and fracture height)
- Borehole and perforation stability
- Sand production and drawdown limits in unconsolidated formations
- Coal evaluation
- Determining the roof-rock-strength index for underground mining operations.

CONTROLLING FACTORS OF METAMORPHIC ROCKS

The world of geology is not a static world. While the rocks we see around us seem stable, everything is undergoing change. Most of the rocks we see are attacked by the forces of weathering, slowly broken down into their mineral components to be formed into new sedimentary rocks.

Weathering is one of the geological processes that help to create new rocks. While weathering works to break down rocks, there is another geologic process that is responsible for creating new rock from existing rock, without the original rock having to be broken down by weathering.

These new rocks are created through increases in heat and/or pressure as the parent rock is buried in the earth. This process is known as metamorphism and creates metamorphic rocks.

As we look at the Rock Cycle we can see that metamorphism involves heat and pressure to change igneous and sedimentary rocks (and other metamorphic rocks) into metamorphic rocks. Metamorphism is defined as the change in the characteristics of a rock in response to changes in temperature, pressure, or fluid content.

Usually these changes do not change the chemical composition of the rock (the minerals remain the same), but the crystallization of new mineral phases. A rock that has been altered through changes in temperature or pressure is called a metamorphic rock. The formation of metamorphic rocks is determined by different controlling factors: composition, temperature and pressure, and foliation.

The composition of a metamorphic rock is determined by the mineral content of the parent rock. Only in very rare instances will a metamorphic rock have a different mineral composition than its parent rock.

Knowing that the composition doesn't change allows a geologist to determine the type of parent rock. Knowing the parent rock allows the geologist to infer much about the origin of the material, allowing for deduction to be made concerning the presence of ancient mountain ranges, coastlines, and seas, even after the sedimentary and igneous rocks have been buried and altered through intense heat and pressure.

Two of the most important controlling factors are temperature and pressure. Though all of the controlling factors work in concert to create metamorphic rocks, temperature and pressure help to create much of the variation.

Temperature affects the changes in a minerals composition, as the temperature increases the mineral is altered. Temperature changes occur as rocks are subducted into the earth, or as hot magma rises from the earth's interior.

Minerals are only stable over a specific range of temperatures that are unique for the mineral. Geologists have done laboratory studies to determine the range of temperatures at which specific minerals are stable. By using this information a geologist can then study a metamorphic rock and determine the temperature of metamorphism at which the rock was formed.

FRACTURE OF BRITTLE ROCKS

Faults - Faults occur when brittle rocks fracture and there is an offset along the fracture. When the offset is small, the displacement can be easily measured, but sometimes the displacement is so large that it is difficult to measure.

Types of Faults

Faults can be divided into several different types depending on the direction of relative displacement. Since faults are planar features, the concept of strike and dip also applies, and thus the strike and dip of a fault plane can be measured. One division of faults is between dip-slip faults, where the displacement is measured along the dip direction of the fault, and strike-slip faults where the displacement is horizontal, parallel to the strike of the fault.

- Dip Slip Faults - Dip slip faults are faults that have an inclined fault plane and along which the relative displacement or offset has occurred along the dip direction. Note that in looking at the displacement on any fault we don't know which side actually moved or if both sides moved, all we can determine is the relative sense of motion.

For any inclined fault plane we define the block above the fault as the *hanging wall block* and the block below the fault as the *footwall block.*

Normal Faults - are faults that result from horizontal tensional stresses in brittle rocks and where the hanging-wall block has moved down relative to the footwall block.

Horsts & Gabens - Due to the tensional stress responsible for normal faults, they often occur in a series, with adjacent faults dipping in opposite directions. In such a case the down-dropped blocks form *grabens* and the uplifted blocks form *horsts*. In areas where tensional stress has recently affected the crust, the grabens may form *rift valleys* and the uplifted horst blocks may form linear mountain ranges. The East African Rift Valley is an example of an area where continental extension has created such a rift. The basin and range province of the western U.S. (Nevada, Utah, and Idaho) is also an area that has recently undergone crustal extension. In the basin and range, the basins are elongated grabens that now form valleys, and the ranges are uplifted horst blocks.

Half-Grabens - A normal fault that has a curved fault plane with the dip decreasing with depth can cause the down-dropped block to rotate. In such a case a half-graben is produced, called such because it is bounded by only one fault instead of the two that form a normal graben.

Reverse Faults - are faults that result from horizontal compressional stresses in brittle rocks, where the hanging-wall block has moved up relative the footwall block.

A *Thrust Fault* is a special case of a reverse fault where the dip of the fault is less than 15°. Thrust faults can have considerable displacement, measuring hundreds of kilometers, and can result in older strata overlying younger strata.

- *Strike Slip Faults* - are faults where the relative motion on the fault has taken place along a horizontal direction. Such faults result from shear stresses acting in the crust. Strike slip faults can be of two varieties, depending on the sense of displacement. To an observer standing on one side of the fault and looking across the fault, if the block on the other side has moved to the left, we say that the fault is a *left-lateral strike-slip fault*. If the block on the other side has moved to the right, we say that the fault is a *right-lateral strike-slip fault*. The famous San Andreas Fault in California is an example of a right-lateral strike-slip fault. Displacements on the San Andreas fault are estimated at over 600 km.

Transform-Faults are a special class of strike-slip faults. These are plate boundaries along which two plates slide past one another in a horizontal manner. The most common type of transform faults occur where oceanic ridges are offset. Note that the transform fault only occurs between the two segments of the ridge.

Outside of this area there is no relative movement because blocks are moving in the same direction. These areas are called fracture zones. The San Andreas fault in California is also a transform fault.

EVIDENCE OF MOVEMENT ON FAULTS

Slikensides are scratch marks that are left on the fault plane as one block moves relative to the other. Slickensides can be used to determine the direction and sense of motion on a fault.

Fault Breccias are crumbled up rocks consisting of angular fragments that were formed as a result of grinding and crushing movement along a fault.

Folding of Ductile Rocks: When rocks deform in a ductile manner, instead of fracturing to form faults, they may bend or fold, and the resulting structures are called *folds*.

Folds result from compressional stresses acting over considerable time. Because the strain rate is low, rocks that we normally consider brittle can behave in a ductile manner resulting in such folds.

We recognize several different kinds of folds.

Monoclines are the simplest types of folds. Monoclines occur when horizontal strata are bent upward so that the two limbs of the fold are still horizontal.

Anticlines are folds where the originally horizontal strata has been folded upward, and the two limbs of the fold dip away from the hinge of the fold.

Synclines are folds where the originally horizontal strata have been folded downward, and the two limbs of the fold dip inward toward the hinge of the fold. Synclines and anticlines usually occur together such that the limb of a syncline is also the limb of an anticline.

- Geometry of Folds - Folds are described by their form and orientation. The sides of a fold are called *limbs*. The limbs intersect at the tightest part of the fold, called the *hinge*. A line connecting all points on the hinge is called the *fold axis*. In the diagrams above, the fold axes are horizontal, but if the fold axis is not horizontal the fold is called a *plunging fold* and the angle that the fold axis makes with a horizontal line is called the *plunge* of the fold. An imaginary plane that includes the fold axis and divides the fold as symmetrically as possible is called the *axial plane* of the fold.

Note that if a plunging fold intersects a horizontal surface, we will see the pattern of the fold on the surface.

Classification of Folds

Folds can be classified based on their appearance.

- o If the two limbs of the fold dip away from the axis with the same angle, the fold is said to be a *symmetrical fold*.
- o If the limbs dip at different angles, the folds are said to be *asymmetrical folds*.
- o If the compressional stresses that cause the folding are intense, the fold can close up and have limbs that are parallel to each other. Such a fold is called an *isoclinal fold* (iso means same, and cline means angle, so isoclinal means the limbs have the same angle). Note the isoclinal fold depicted in the diagram below is also a symmetrical fold.
- o If the folding is so intense that the strata on one limb of the fold becomes nearly upside down, the fold is called an *overturned fold*.
- o An overturned fold with an axial plane that is nearly horizontal is called a *recumbant fold*.
- o A fold that has no curvature in its hinge and straight-sided limbs that form a zigzag pattern is called a *chevron fold*.

The Relationship Between Folding and Faulting

Because different rocks behave differently under stress, we expect that some rocks when subjected to the same stress will fracture or fault, while others will fold. When such contrasting rocks occur in the same area, such as ductile rocks overlying brittle rocks, the brittle rocks may fault and the ductile rocks may bend or fold over the fault. Also since even ductile rocks can eventually fracture under high stress, rocks may fold up to a certain point then fracture to form a fault.

Folds and Topography

Since different rocks have different resistance to erosion and weathering, erosion of folded areas can lead to a topography that reflects the folding. Resistant strata would form ridges that have the same form as the folds, while less resistant strata will form valleys.

Mountain Ranges - The Result of Deformation of the Crust

One of the most spectacular results of deformation acting within the crust of the Earth is the formation of mountain ranges. Mountains originate by three processes, two of which are directly related to deformation. Thus, there are three types of mountains:

1. *Fault Block Mountains* - As the name implies, fault block mountains originate by faulting. Both normal and reverse faults can cause the uplift of blocks of crustal rocks. The Sierra Nevada mountains of California, and the mountains in the Basin and Range province of the western U.S., as discussed previously, were formed by faulting processes and are thus fault block mountains.
2. *Fold & Thrust Mountains* - Large compressional stresses can be generated in the crust by tectonic forces that cause continental crustal areas to collide. When this occurs the rocks between the two continental blocks become folded and faulted under compressional stresses and are pushed upward to form fold and thrust mountains. The Himalayan Mountains (currently the highest on Earth) are mountains of this type and were formed as a result of the Indian Plate colliding with the Eurasian plate. Similarly the Appalachian Mountains of North America and the Alps of Europe were formed by such processes.
3. *Volcanic Mountains* - The third type of mountains, volcanic mountains, are not formed by deformational processes, but instead by the outpouring of magma onto the surface of the Earth. The Cascade Mountains of the western U.S., and of course the mountains of the Hawaiian Islands and Iceland are volcanic mountains.

TRACE ELEMENTS

As mentioned before, the trace element concentrations found in subduction related volcanic rocks are not consistent with derivation of the magmas from partial melting of subducted oceanic crust. First we will look at what the trace element concentrations show, then discuss why they are not consistent with an origin involving direct melting of the oceanic lithosphere, and then discuss how the trace element patterns of subduction related magmas might develop. The subducted oceanic lithosphere likely does make a contribution to calc-alkaline magmas, but not necessarily as the primary source of these magmas.

When plotted on an incompatible trace element, calc-alkaline rocks show an irregular pattern with many peaks and valleys, unlike the relatively smooth patterns exhibited by OIBs and MORBs. First, in calc-alkaline basalts (as well as andesites, not shown) the heavy REE, represented by Yb are not depleted relative to MORBs.

Second, the elements Nb, Ta, and Ti show negative anomalies (depletion) relative to elements like Ba, K, La, and Ce. We'll discuss the implications of each of these points below:

1. Lack of HREE depletion. When oceanic crust is subducted it traverses a path of increasing pressure and temperature. In doing so it is metamorphosed. The mineral assemblage of olivine, pyroxene and plagioclase is not stable at higher pressures and temperatures and is metamorphosed into an assemblage of mostly pyroxene and garnet. Such a rock is called an*eclogite*. At the temperatures and pressures expected to be present at the depth of the subduction zone below the volcanic arc the basalt would in fact be an eclogite, yet its chemical composition will not have changed, i.e. it will still show a MORB-like trace element pattern.

   Heavy REEs are compatible in garnet. Thus if garnet is present in the source, HREEs will be held back by the garnet during partial melting, and thus any liquids produced should show lower HREE concentrations than the original source. Thus, if the source of calc-alkaline magmas is postulated to be oceanic crust (now an eclogite) we would expect to see HREE concentrations lower than oceanic crust. Since this is not observed, it is unlikely that the calc-alkaline magmas represent partial melts of the subducted oceanic crust.
2. Depletion in Nb, Ta, and Ti relative to K, La, and Ce. The elements Nb, Ta, and Ti are elements called *High Field Strength Elements (HFSE)*. These are elements that have a small ionic radius and a high charge (usually +4). Because of their high charge and small size they are not readily soluble in aqueous fluids. During chemical weathering or dehydration for example these elements remain in the solids. Elements like Ba and K are often called *Large Ion Lithophile Elements (LILE)*. The LILE and REE (like La and Ce) are highly soluble in aqueous fluids. Thus, dehydration of the subducted oceanic crust would be expected to release fluids that have high concentrations of LILE and REE and low concentrations of HFSE. If these fluids then interacted with the

mantle overlying the subducted plate they would change the composition of that mantle (metasomatize it) so that it would have a trace element pattern more similar to the pattern observed in calc-alkaline volcanic rocks. Subsequent melting of this metasomatized mantle would explain the trace element pattern observed in the erupted magmas.

The addition of fluids has an additional effect, in lowering the solidus temperature of the mantle and inducing partial melting. Furthermore, addition of fluids at the source of calc-alkaline magmas would explain why these magmas would become saturated with water at lower pressures to fractionate to produce the calc-alkaline trend on an AFM.

Isotopes

Sr and Nd isotopic ratios for subduction-related volcanic rocks are similar to OIBs, but show higher ratios of $^{87}Sr/^{86}Sr$, and extend to lower ratios of $^{143}Nd/^{144}Nd$. Three points are notable about this data.

The subduction related rocks do not generally show Sr and Nd isotopic ratios similar to MORBs. This suggests that they were not derived from partial melting of subducted oceanic crust or from partial melting of an unmodified MORB source.

The offset from OIBs to higher values of $^{87}Sr/^{86}Sr$ at constant $^{143}Nd/^{144}Nd$ could be explained by addition of seawater to the source of the subduction-related rocks. Seawater has relatively high concentrations of Sr and extremely low concentrations of Nd. Thus, if seawater expelled from the subducted lithosphere were incorporated into the mantle source, it would raise the $^{87}Sr/^{86}Sr$ ratio and have little effect on the $^{143}Nd/^{144}Nd$ ratio.

The extension of the array toward higher $^{87}Sr/^{86}Sr$ and lower $^{143}Nd/^{144}Nd$ suggests that a continental crustal component is incorporated into subduction related magmas. This could come from either subducted oceanic sediments or crustal contamination. If sediments are subducted, they would metamorphose along with the rest of the oceanic crust and upon dehydration the fluids could carry an isotopic signature contributed by the sediments. In general, rocks erupted in continental settings show higher $^{87}Sr/^{86}Sr$ and lower $^{143}Nd/^{144}Nd$ ratios than those erupted in island arcs. This argues for some contamination of the magmas by the crustal rocks through which the magmas pass.

The extent of subducted sediment involvement can, in some cases be evaluated by looking at an isotope of Beryllium, ^{10}Be, and a highly incompatible trace element Boron (B).

^{10}Be is an isotope of Be that is produced in the upper atmosphere by bombardment of cosmic rays. Once produced it has a half-life of 1.5 million years. Be and its radioactive isotope are absorbed by the oceans and are adsorbed onto the surface of clay minerals. Because of its short half life, only small quantities ^{10}Be remain after the passage of more than about 10 million years. If oceanic sediment is subducted and contributes material to the source of calc-alkaline magmas before it has completely decayed, then we should see small amounts of ^{10}Be in the magmas and rocks. But note that this will only be true for very young sediment and in young volcanic rocks. So, we do not necessarily expect to see ^{10}Be in all subduction related rocks.

B is an element that is abundant in sediments but has very low concentrations in the mantle. B/Be ratios for mantle rocks are also low. Furthermore, B is much more readily soluble in fluids, so fluids derived from dehydration of the sediments have much higher B/Be ratios than the mantle and sediments. Thus, if we look at the concentration of ^{10}Be and the B/Be ratio in subduction-related volcanic rocks we may be able to determine whether or not sediments and fluids are involved in the production of subduction-related magmas.

What is found is that for some arcs the concentration of^{10}Be increases linearly with B/Be ratio. Some arcs like the Central American Arc, the Kurile arc, and the New Britain Arc range to high values of ^{10}Be indicating the incorporation of some component of relatively young oceanic sediment. Other arcs like the Aleutians, Chile, and Kamchatka have low ^{10}Be, but extend to higher values of B/Be. Note that fluids could have rather variable ^{10}Be and B/Be, but the data clearly indicates that some fluid is contributing to the source of subduction-related magmas, consistent with other information we have examined.

Origin

Over the last fifteen years or so we have come to a much clearer picture of the origin of the calc-alkaline suite of rocks that is commonly associated with subduction. We here summarize the current theory based on the information discussed above:

- Subduction carries oceanic crust and sediment to depth. As the pressure and temperature rise, the MORB crust and sediments undergo metamorphism that releases hydrous fluids.
- These hydrous fluids carry with them high concentrations of LILE and REE, but leave behind the relatively insoluble HFSE. They also carry the isotopic signature of the basaltic crust sediment mixture that released the fluids, and thus have higher ^{87}Sr/^{86}Sr ratios and lower^{143}Nd/^{144}Nd ratios, reflecting the isotopic composition of the subducted material. If the sediments are young, they may contribute ^{10}Be to these fluids.

- The fluids act to metasomatize the overlying mantle wedge, enriching it in LILE, REE, B,$^{87}Sr/^{86}Sr$, and possibly ^{10}Be and lowering the $^{143}Nd/^{144}Nd$ ratio of this mantle.
- Adding H_2O to the mantle wedge lowers the solidus temperature allowing for partial melting of this metasomatized mantle and generating hydrous basaltic magmas.
- This hydrous basaltic magmas become saturated with water at crustal depths and differentiate by crystal fractionation, possibly accompanied by contamination of crustal material, to generate the andesites, dacites, and rhyolites of the calc-alkaline suite.

8

Climatic and Environmental Change

In the absence of detailed regional or location specific palaeoclimatic data, this discussion draws on information from the broader Indo-Pacific region to provide the context for this chapter. It is widely acknowledged that there has been an increase in climatic variability in the Australasian region in the last few thousand years, in particular from approximately 2000 BP to the present. Many of the longer-term trends in climate change that have occurred during the period spanning the mid Holocene to the present day are related to the El Nino/Southern Oscillation (ENSO) cycle, which has strongly influenced climatic patterns in Australia, and at present represents the principal source of inter-annual climatic variability within the Indo-Pacific region. Sea level changes too are inter-woven with long-term climatic shifts, particularly the intensity of the summer monsoon and cycling periods of aridity and increased precipitation linked to the ENSO cycle.

This is due to the proximity of north Australia to the Western Pacific Warm Pool, which is responsible for the largest transfer of heat from the Pacific Ocean into the Indian Ocean and is implicated in the generation of El Nino/La Nina phases of the southern oscillation. Within these longer-term environmental patterns, there appear to be several punctuated phases of relatively rapid climatic change throughout the Holocene, as well as higher levels of aridity in tropical areas. Most of these climate change events are characterised by polar cooling, tropical aridity and major atmospheric circulation changes, events that are seen to occur during the mid to late Holocene in regional Australasian tropical palaeoclimate records.

While the long-term climate change patterns are important for contextualising the degree of variability in climate history throughout the Holocene, possibly the most significant period of climatic change for our purposes relates to the last 1200 years. In the northern hemisphere, there is evidence for a warm and dry period about between 1200 to 700 BP in low latitudes named the "Medieval Warm Period" or "Little Climatic Optimum" (LCO). Similarly, there is evidence for a cool dry period following the LCO between approximately 600 to 100 BP (Jones et al. 2001), referred to as the "Little Ice Age" (LIA). Based on glacial advances in both hemispheres and enhanced polar atmospheric circulation, it is strongly suggested that the LIA was a global scale event. However, the timing and nature of these significant events is not well established for the Southern Hemisphere, especially for the tropics.

The LIA and LCO and long-term El Nino events are interrelated, and have had strong ecological and economic consequences. The best-studied impacts of these processes on ecosystems are in marine environments, where El Nino is correlated with dramatic changes in the abundance and distribution of many organisms, and the collapse of fisheries. There are also documented impacts on terrestrial organisms as well, linked to effects on the structure of the vegetation. For example, El Nino events have been linked to the almost complete defoliation of mangrove forests. Therefore, ENSO-related climatic oscillations must have had significant impacts on human populations, primarily affecting the resource base in terms of the long and short-term availability, stability and structuring of these resources.

In addition to the effects of climatic events, the resource base for past hunter-gatherers would have been influenced by processes of landscape alteration linked to climatic and sea level changes, which have created drastically different environments over time in Arnhem Land coastal areas. Following the cessation of sea-level rise c. 6000 BP, as extensive coastal plains developed through sedimentation and coastal progradation, the mangroves that had invaded drowned river valleys (the 'Big Swamp Phase') retreated seawards and had mainly disappeared from the floodplains by 4000 years BP. There followed a so-called Sinuous Phase, during which meandering palaeochannels were created across the plains, resulting in a mosaic of estuarine, freshwater and mudflat areas.

Continued sedimentation and the slowing of coastal progradation during the Cuspate Phase, c. 2500 years ago, led to a cut-off of the tidal influence of the rivers. Freshwater ponded behind the cheniers and in palaeochannels creating the vast freshwater floodplains and wetlands that are a major feature of the northern coastal plains today. Coastal processes, of progradation by marine mud accretion and subsequent colonisation by mangroves may be indirectly related to climatic conditions. As explained by Woodroffe et al., where macrotidal estuaries infill with deposition of tidal sediment from seaward, but where catchment areas are small, wet season flooding may not be large enough to reopen channels and move sediment downstream again.

In other words, these estuarine plains develop through mainly marine deposition within broadscale low input from rivers in the wet season. Thus a marine rather than fluvial source for infill and muddy progradation is consistent with a long-term adds period. There does appear to be a general regional trend towards increased aridity in the mid to late Holocene, as supported by data extracted from coral, foraminifera, varve, lake and sea bottom sediments from sites in the Australian and circum-Pacific region. The pattern of climatic shifts observed in recent reviews of the climate history of the south Pacific region also indicates a change from low seasonality in the early Holocene to increased seasonality in the late Holocene (Markgraf et al. 1992; Shulmeister 1999:86). Evidence from pollen records on Groote Eylandt are indicative of this environmental change from continuously increasing rainfall (effective precipitation) during the early Holocene, to a period of reduced rainfall and increased climatic variability after 4000 BP.

On the Groote Eylandt archipelago, effective precipitation declined sharply soon after 3700 BP, with evidence across Australia indicating that climate became more variable after this time. Geomorphic data based on the dating of cheniers and on sediment and pollen records from cheniers and coastal dunefields indicates that some of the observed changes in these systems are synchronous across north Australia, and may represent coherent, broad-scale climatic signals. It is argued by Shulmeister (1992:113) that the widespread occurrence of dune systems during the mid to late Holocene period, together with pollen data from the lake deposits on Groote Eylandt, suggests a climatically driven process characterised by increasing aridity and climatic variability across north Australia due to the onset of ENSO conditions similar to that of the present-day.

Consistent with a more add period of reduced wet season precipitation, interspersed with sudden episodes of high rainfall are sedimentary records of sporadic major floods in north Australian river systems during the late Holocene. Most of the environmental studies in this region tend to focus on the mid Holocene. However, two palaeoclimate studies at Groote Eylandt and Vanderlin Island off the east Arnhem Land coast do provide some data for the Late Holocene period of the last 2000 years. Pollen records from the Four Mile Billabong and Walala lake on each of these islands respectively, show peaks for this period in dryland taxa including Pandanus and Sapindaceae, which indicate expansion of more open woodland and are noted as markers of increased burning and disturbed environments.

The diversity of diatoms also decrease in this period in the Walala record, consistent with an overall aridity trend and increased climate variability for the period around the LCO and LIA. In order to explore the relationship between climate change, recorded at a broad-scale level, and human behaviour in the late Holocene, we present an overview through three case studies, in the form of a finer scaled archaeological record of changes in hunter-gatherer economic and social practice.

Hunter-gatherer Behaviour in the Arnhem Land Region

Archaeological evidence of changes in hunter-gatherer behaviour have been recorded from several archaeological sites across the Arnhem Land coastal region; on the Beagle Gulf mainland at the western end of Arnhem Land, in central Arnhem Land and in northeastern Arnhem Land. Each of the study areas lie within the latitudes of 11° and 14°s, along the low-energy north Australia shoreline, mostly of Holocene age, that is still prograding, through deposition of both seaward and terrestrial sedimentation. Geomorphic studies show that these cultural landscapes are less than 3000 years old. Earlier deposits may have been removed by sea level oscillations at around 3000-3500 years BP, as suggested by a gap in stratigraphy for this period on the coastal plains of the Beagle Gulf and neighbouring Van Diemen Gulf shorelines.

Hope Inlet, Beagle Gulf, western end of Arnhem Land

Hope Inlet is a small estuary facing the Beagle Gulf, at the western end of Arnhem Land, and about 25 km north-east of Darwin. The Hope Inlet landscape is at present dominated by coastal chenier plains, which are fringed at the seaward side by mangrove forests hundreds of metres wide. In this less than 100 [km.sup.2] of the coast, comprising a series of headlands protruding onto salt flats, each partitioned by tidal channels, mangroves and swamps, are over 200 Aboriginal shell middens and mounds, earth mounds and shell and exotic stone artefact scatters.

Archaeological research in the mid 1990s (Bourke 2000; 2004) revealed the temporal relationship of three excavated shell mounds—HI81, HI83 and HI80—located within a few hundred metres of each other in adjacent microenvironments on one headland and associated salt fiats area. Two of the mounds (HI81 and HI83), located on the seaward open headland margins and in woodland 300 m inland respectively, overlap in their period of use between 2000 and 1400 calBP. The third mound (HI80) formed on nearby salt fiats centuries later within the period 1000 to 500 calBP. Faunal remains indicate a focus on near-shore marine and estuarine food sources; fish, crab and shellfish and small proportions of bird and terrestrial mammals and reptiles.

A predominance of Anadara granosa in the molluscan remains (HI81 - 97%; HI83 - 92%; HI80 - 72%) - a bivalve that inhabits open mudflats - is typical of hundreds of this type of tropical shell mound across the north Australian coast. Also common in small quantities to these Late Holocene mounds is a suite of gastropods from the mangroves, usually Cassidula angulata, Nerita spp. Terebralia semistriata, Telescopium telescopium, Ellobium aurisjudae and Chicoreus capucinus. Between the periods 2000-1400 calBP and 1000-500 calBP a marked change occurs in the molluscan assemblage, represented by a 20% decrease in frequency of the dominant species Anadara granosa, and a corresponding increase in mangrove gastropod species. This trend is ongoing after the mound-building period.

On the salt fiats surface are scatters of shell and stone artefacts dominated by mangrove gastropod species Terebralia and Telescopium. These scatters can be placed within the period 500-200 calBP on geomorphic and archaeological grounds. Today in this section of the coast Anadara shell beds are rare. The most common large edible molluscs are gastropods Telescopium, Terebralia and Nerita spp., which live within the mangrove forests that dominate the present day shoreline. Thus for the period represented in the Hope Inlet landscape, spanning 2000 to 500 calBP, the shell deposits provide an archaeological record of gradual change in the focus of mollusc taxa targeted by the Late Holocene inhabitants. The observed trend culminates in a marked change in cultural behaviour, from mound-building to a shell discard practice that results in low, horizontally spread out shell middens as observed ethnographically.

Blyth River, Central Arnhem Land

The Blyth River is one of several large rivers that drain the central northern coast of Arnhem Land. It rises in the sandstone country of the Arnhem Land plateau and flows through Tertiary laterite plains and open eucalypt woodland before crossing the recently formed coastal plains and dumping into the Arafura Sea. Sandy beaches stretch away east and west for 60 km. The fiver is tidal up to 18 km from its mouth and its banks along this stretch are lined with thick mangrove forest. The coastal plains of the Blyth River contain diverse vegetation systems, including relict fiver channels on the floodplains that form backwater swamps supporting freshwater vegetation, and chenier dune systems that support monsoon rainforest pockets and pandanus scrub.

The chenier adjacent to the coast has been dated to 1400 BE The next chenier inland has been dated to 2000 BP. Evidence of human occupation of the Blyth River region is provided by numerous sites, predominantly shell mounds and middens, dotted along the coast and inland. The oldest site so far in the area is Maganbal, a midden nowadays located 10 km from the coast and dated to c. 3400 calBE However, the majority of sites occur closer to or on the coast and are considerably younger, dated to between 1200 calBP and modem, reflecting the recent nature of this landscape.

Ji-bena, an earth mound bordering a large freshwater swamp close to the mouth of the Blyth River, was occupied initially approximately 1000 calBP. The shell analysis reveals that between approximately 1000 and 950 calBP, there was a dramatic increase in the presence of Dosinia sp. Dosinia are rapid, deep burrowing bivalves that occur in open areas of sand and silt with good current flow. They are reported by Meehan to be collected from the mid-littoral zone. Dosinia continued to be the dominant molluscan species exploited at the site until approximately 800 years BP when they disappeared from the archaeological record. This event is also reflected in other nearby sites.

Post 800 years calBP, the economy at Ji-bena focused increasingly on freshwater resources, although marine shellfish continued to be taken at a much lower level. Mangrove species were continually exploited throughout the history of the site, albeit to a reduced level after 800 calBP. These cultural changes appear to correlate with the transition from estuarine to freshwater conditions in the neighbouring swamp, marked by a reorganisation in foraging relative to available resources.

Blue Mud Bay, Eastern Arnhem Land

Archaeological research on Point Blane peninsula, the central of three peninsulas located within the coastal plains on the northem coastal margins of Blue Mud Bay in northeast Arnhem Land, provides further supporting evidence for correlations between human behaviour and environmental changes in the late Holocene. Radiocarbon age determinations indicate that the sites located in the area fall within approximately 3000 calibrated years BP and the present. Based on these chronological patterns and variability in site distribution, morphology and shellfish species representation, several phases of occupation and resource use within the area have been identified. There appears to have been limited or low level use of this area prior to 2500 calBP, although this may relate to factors of site visibility and/or post-depositional destruction of archaeological material. Between approximately 2500 and 500 calBP, there was a phase of intensive occupation on the margin of the present-day wetlands (Grindall Bay), corresponding with a large number of Anadara granosa dominated shell mound and midden deposits.

In contrast to these patterns of relatively intensive occupation and resource exploitation, based on the paucity of archaeological evidence on the exposed coastal margin (Myaoola Bay) during this period, occupation in this area of the Point Blane peninsula is of a relatively low intensity until approximately 1000 calBP. With the cessation of Anadara granosa dominated mound accumulation on the peninsula occurring at around 500 calBP, there is a corresponding increase in the evidence for occupation and resource use in Myaoola Bay. Related to these chronological patterns of occupation, the overall pattern of molluscan resource and habitat use on the Point Blane peninsula is as follows: prior to approximately 2500 calBP there was a greater focus on the shallow water, near shore zone in conjunction with the sand and mud flats.

Associated with the period of mound formation in the study area, between approximately 2500 and 500 calBP there was a decline in the relative abundance of species from the shallow water, hard substrate areas and an increasingly heavy reliance on species from the sand and mud flats, particularly Anadara granosa and Mactra abbreviata, with some variability in the use of species from the mangroves through time depending on site location. After approximately 500 calBP, there was a 35% increase in the use of mangrove species, such as Isognomon isognomon, corresponding with a decline of approximately 38% in the frequency of sand and mud flat bivalves.

These contrasting chronological patterns of exploitation directly reflect the dynamic and changing nature of the coastline throughout the late Holocene. Several phases of economic reorganisation have also been identified during this time, which correspond with identifiable phases of climatic change and the differential availability of resources, a process that has been interpreted as resulting in changes to both logistical and residential mobility. This is particularly evident in the shift from shell mound formation at approximately 500 calBP on Grindall Bay to the deposition of low-lying, horizontally spread midden sites on Myaoola Bay, a pattern that is very similar to that noted above in Hope Inlet.

Although the exact nature of the LIA in the Pacific is uncertain, these environmental effects can be indirectly related to an overall aridity trend, as identified by some Australian researchers. Moreover there is sufficient overlap between the phases of climate change previously described, particularly the Little Climatic Optimum (1200 to 700 BP) and the Little Ice Age (600 to 100 BP), and the timing of behavioural changes within each of the three regions discussed to suggest that, with regional variation in the nature and severity of the changes, there was an associated human response to late Holocene climatic variability. It should be stressed that these are not responses to catastrophic events, but rather behavioural changes relative to a complex pattern of inter-related environmental and climatic episodes that culminated broadly with the LCO and LIA.

The patterns of change in foraging and cultural disposal practices highlighted here are not confined to Hope Inlet, the Blyth River and the Point Blane peninsula. The broader regional extent of the patterns seen here is particularly evident when comparing the chronological evidence for mounding behaviour and/or variability in the exploitation of particular molluscan species in coastal areas across northern Australia. Radiocarbon dates for the occurrence of shell mounds across the north generally fall within the period spanning approximately 3000 BP and 500 calBP, with corroborating evidence in other studies in western Arnhem Land (Mowat 1995) and Cobourg Peninsula, and right across northern Australia, from the Kimberly region to the north Queensland region at Bayley Point, Weipa and Princess Charlotte Bay. That similar patterns of foraging reorganisation were occurring within this period, in areas that are separated by large distances but within a restricted latitude, indicates that broader scale processes of environmental change were indeed the primary cause behind economic change during the late Holocene.

This is indicated by changes to the faunal assemblages that reflect habitat shifts at specific times, especially around 2500 calBP and between 800 and 500 BP, in each of the three study areas discussed. Potentially the strongest indication of the link between climatic conditions and environmental effects relates to the effective removal of suitable habitats for mudflat bivalves. This event occurred with the transition to the LIA between 800 to 600 years ago, depending on regional variation in the timing of these changes. This is particularly evident for Anadara granosa, a species abundantly represented in shell mounds, that has either disappeared entirely, or occurs today only in very low densities in northern Australia. The disappearance of this species appears to be associated with the cessation of the cultural practice of mound building across a number of north Australian coastal areas, such as Darwin Harbour, the Point Blane peninsula, Weipa and Princess Charlotte Bay.

Mudflat bivalves are known to have specific habitat requirements. Anadara granosa for example, are essentially soft substrate dwellers, which are intertidal or marginally subtidal in distribution, and require a number of optimal, environmental conditions for shell bed establishment and proliferation. This species occurs naturally in large tropical estuarine mudflats situated outside the mouth of estuaries and tidal creeks, bordered on the landward margin by mangrove forests. It thrives in high densities under comparatively calm conditions especially in shallow inlets or bays, with a sub-stratum of fine, soft, flocculent mud. These optimal habitats are protected from strong wave action, although the current should be strong enough to transport natural food. The three most important ecological criteria for the establishment and proliferation of Anadara shell beds are the nature of the substrate, salinity levels (ranging between 18 to 30 parts per thousand) and a seaward slope of bed of 5-15°.

It can be seen that there are a number of interrelated environmental and climatic factors that contributed to the disappearance of the habitats and conditions suitable for Anadara granosa biomass around 800 to 500 BP. Continuing processes of sedimentation would have gradually changed the gradient of the coastal plain and slope of the shore, eventually reducing tidal inundation and freshwater input, leading to changing conditions for Anadara. An additional, related factor that may explain the decline of Anadara is the sustained accretion of the mudflats and subsequent colonisation by mangroves with the elevation of the substratum and gradual isolation of the shell beds from the tides. There also appear to have been changes in sea levels and salinity levels at this time that would have had a significant impact on the large Anadara beds.

A change in the archaeological record from dominance by mudflat bivalves to mangrove gastropods also has environmental implications due to the biological differences of these taxa. The Potamididae and Ellobididiae mangrove gastropods Telescopium, Terebralia, Ellobium and Cassidula are euryhaline. That is, they are much more tolerant of long temporal extremes and wide fluctuations in salinity and temperature than Arcidae mudflat bivalves such as Anadara, which are stenohaline. There is a strong possibility that similar factors may have contributed to the disappearance of the sandy/mudflat bivalve Dosinia sp. from the Ji-bena mound on the Blyth River around 800 calBP, whereas mangrove species continued to persist.

In north Australia, the changes seen in these case studies support the argument that Aboriginal mollusc exploitation reflects change in local ecological habitats, reflecting those broader coastal environmental changes that occurred over the last few thousand years. While more research is required into dating geomorphic changes in local environments and in biological studies of mudflat ecology, there is increasing evidence for a change to more arid and variable climatic conditions in late Holocene north Australia that would have implications for foraging peoples.

For example, a number of researchers have highlighted broader patterns of coastal change characterised by an increasing aridity and northward movement of the northwest Australian monsoon. In addition, episodic, unpredictable environmental phenomena, such as torrential rains, flooding and droughts, associated with ENSO events, are known to disrupt both marine and terrestrial habitats and therefore resource availability and predictability.

Further evidence comes from recent research in the Torres Strait, where patterns evident within the archaeological record indicate relative stability of both economic and social behaviour until about 800 to 600 years ago, and again between approximately 500 to 400 years ago. McNiven summarises archaeological evidence from across Torres Strait that demonstrates broad-scale synchronous cultural change in settlement patterns, demography, mobility, rituals, seascape construction, social alliances and exchange relationships between 800 and 600 years ago. As with the argument presented here, it is suggested that people were responding to significant climatic shifts during this period, particularly between 800 and 600 years ago, and possibly during the more recent phase.

Within the Torres Strait region it has also been noted that the sand and mud flat species Anadara antiquata dominated midden assemblages in areas where these habitats no longer exist (Barham et al. 2004:33), parallelling those patterns noted in the examples presented above. Further, McNiven (2006:9), quoting Spriggs (1993:198), draws attention to widespread changes in settlement patterns and material culture in island Melanesia dated from c. 700 years ago. This period also witnessed the appearance of agricultural developments such as raised bed swamp cultivation at Kuk and the adoption of the agroforestry trees Casuarina in New Guinea and Alnus by Inca peoples in the Peruvian Andes.

While extreme climatic events may impact severely on agricultural societies, hunter-gatherer people are known to be more adaptable to significant climate change due to their flexible, broad spectrum foraging strategies for dealing with both temporary and long-term unavailability of particular resources (Meehan 1982) and other risk-reducing strategies such as mobility and exchange (Thomson 1949). The case studies cited above indicate that foraging behaviour on the coastal margins of northern Australia was generally flexible, and that people actively altered their foraging strategies to incorporate newly available or increasingly abundant species within the general pattern of environmental and climatic change previously outlined. However, hunter-gatherer adaptive strategies do have climatic thresholds of cultural tolerance that may be triggered by extended periods of unpredictable climatic events.

In the north Australian region, major cultural changes shown in the archaeological record, such as the end of shell mounding a cultural practice that had persisted for millennia—may have been triggered by a coincidence of particular historic environmental and social events.

One salient social event that would correspond to the transitional LCO to LIA period if the (previously discounted) 1100 AD radiocarbon dates for this event are corroborated by future research, is the arrival of Macassans on the Arnhem Land coast. Despite hunter-gatherer stability and resilience in the face of climatic variability, there is evidence elsewhere that significant climatic transitions are marked in the archaeological record, such as the first construction of large temple mounds along the Peruvian coast with the mid-Holocene onset of ENSO.

During the mid to late Holocene, distinct cultural changes have also been observed in many other areas of the world, including population growth, population dislocations and changes in settlement patterns, urban abandonment, state collapse, increasing conflict and the cessation of long-distance trade.

In the Pacific region, the environmental and cultural effects of climate change in the last 1000 years have been the topic of extensive recent debate.

Between 900 and 300 BP, numerous fortified settlements were built throughout island Southeast Asia, Oceania and the Pacific. Previously, many archaeologists have explained these phenomena as a result of conflict arising from increased population, more competition for resources and the development of social complexity. More recently however, archaeologists working in Fiji have related fort-building and demographic change with the transition between the LCO and LIA. However, direct physical links between climatic and environmental change and the archaeological record have yet to be demonstrated unambiguously. The case studies cited above have established that the interpretation of archaeological evidence together with the environmental evidence can be used to support the argument for climate change influencing cultural change in northern Australia. To investigate the patterns noted here further, we require more fine-grained temporal and palaeoclimatic data to determine accurately whether or not environmental and climatic shifts were responsible for the kinds of behavioural changes we are seeing in the archaeological record. Analysis of shellfish exploitation in the archaeological record of other sites in the Pacific region may reveal similar trends and provide a more direct link that is currently missing between changes in climate, environment and human responses over the last millennium.

ENVIRONMENTAL CHANGE AND BIOLOGICAL CONTROLS

Today the activities of one species, humans, are reducing the diversity of all others and transforming the global environment. Ecosystems subjected to the stresses of "global change" (including climate change and altered weather patterns, the depletion of stratospheric ozone, deforestation, coastal pollution, and marked reductions of biological diversity) become more susceptible to the emergence, invasion, and spread of opportunistic species.

When subject to multiple stresses, natural environments can exhibit symptoms that indicate reductions in resilience, resistance, and regenerative capabilities. Conversely, ecosystems have inherent flexibilities and survival strategies that can be strengthened by systematic stress, such as the seasonal battering they must endure in temperate latitudes. But their tolerance for abuse has its limits. Several features of global change tend to reduce predators disproportionately, and in the process release prey from their biological controls. Among the most widespread are:

- Fragmentation and loss of habitat
- Dominance of monocultures in agriculture and aquaculture
- Excessive use of toxic chemicals
- Increased ultraviolet radiation, and
- Climate change and weather instability.

The breaking up of large tracts of forest or other natural wilderness into smaller and more diverse patches reduces the available habitat for large predators, and favours many pests. Land and climate changes may act synergistically, as when constricted habitat frustrates a species' ability to migrate north or south to survive altered climatic conditions. Extensive deforestation and climate anomalies—such as the delayed monsoon rains that resulted from this year's El Nino—can also act synergistically, with costly results. A ready example is the massive haze from burning that covered much of Southeast Asia in September and October, causing acute and chronic respiratory damage and losses in trade, investment, and tourism—the latter, a $26 billion a year industry.

The dedication of land to *monoculture,* that is, the cultivation of single crops with restricted genetic and species diversity, renders plants more vulnerable to disease. Simplified systems are also more susceptible to climatic extremes and to outbreaks of pests.

Over-use of pesticides kills birds and beneficial insects, as noted in 1962 by Rachel Carson. The title of her book, Silent Spring, made reference to the absence of the chorus of birds in springtime, and the resulting resurgence of plant-eating insects—that had also evolved a resistance to pesticides. The worldwide response to her message transformed agricultural policies and generated more enlightened pest management. But today, the heavy application of pesticides still carries risks to both human health and natural systems. Over-use of pesticides in Texas and Alabama to control the boll weevil has alarmed farmers, for friendly insects such as spiders and lady bugs have died off and other plant pests have rebounded.

Ecosystem Health

As noted earlier, one of "nature's services" is to keep opportunistic species under control. Maintaining this service entails sustaining the health and integrity of ecosystems. One of the essentials is genetic and species biodiversity to provide alternative hosts for disease organisms. Another is sufficient stability among functional groups of species (such as recyclers, scavengers, predators, competitors, and prey) to ensure the suppression of opportunists and preserve essential ecological functions. Habitat is crucial.

Stands of trees interspersed with agricultural fields, for instance, support birds that control insects; clean ponds with healthy populations of fish serve to control mosquito larvae; and adequate wetlands filter excess nutrients, harmful chemicals, and microorganisms.

As a case in point, in tidewater Maryland, buffer zones around farms and the restoration of wetlands and river-bed trees can absorb the flow of sediments, chemicals, and harmful organisms into Chesapeake Bay, and thus reduce the emergence and spread of algae, toxic to fish. Ecosystems are also interrelated: healthy forests and mangroves in Central America, for example, are crucial to coral reefs that spawn fish stocks, formed at the origin of the great Gulf Stream.

Maintaining the integrity of natural environmental systems provides generalized defenses against the proliferation of opportunistic pests and disease.

Population explosions of nuisance organisms, be they animals or plants or microbes, often reflect failing ecosystem health: a sign of systems out of equilibrium, in terms of the balance of organisms required to perform essential functions. The damage done, moreover, can be cumulative, for multiply-stressed systems are less able to resist and rebound when other stresses come along.

Rodents, insects, and algae are thus key biological indicators of ecosystem health. Their populations and species compositions respond rapidly to environmental change—particularly to an increase in their food supply, or a drop in the number of their natural predators. These indicator species are also linked to human health.

Impacts of a Loss of Biodiversity

The present rate of species extinctions around the world is a potential threat to human health when one considers the role that predators play in containing infectious disease. From the largest to the smallest scales, an essential element in natural systems for countering stress is a diversity of defenses and responses. Thus, animals that seem redundant may serve as "insurance" species in a natural ecosystem, providing a back-up layer of resilience and resistance when others are lost from disease, a changing environment, or a shortage of food or water.

In 1996 the World Conservation Union reported that one-fourth of all species of mammals—and similar proportions of reptiles, amphibians, and fish—are threatened. The current rate of extinctions (estimated at 100 to 1,000 times the rate of loss in the pre-human era) falls heaviest on large predators and "specialists," and thus may initially favour the spread of opportunistic species.

ENVIRONMENTAL IMPACT ASSESSMENT OF FORESTRY PROJECTS

The current Regulations are The Environmental Impact Assessment (Forestry) (England and Wales) Regulations 1999 [SI 1999/2228] and the Environmental Impact Assessment (Forestry) (Scotland) Regulations 1999 [SI 1999/43]. These Regulations have been further amended by the Environmental

Impact Assessment (Forestry) (England and Wales) (Amendment) Regulations 2006 and The Environmental Impact Assessment (Scotland) Amendment Regulations 2006. These Regulations came into force on 6th September 1999 and require anyone who wishes to carry out a relevant project (i.e. deforestation, afforestation, forestry roads or quarries that might have a significant effect on the environment) to obtain consent for the work from the Forestry Commission. The applicant or proposer must submit an Environmental Statement in support of the proposals to apply for consent.

You will find below some of the terms used in the EIA Regulations:

a) Appropriate Authority-The Secretary of State for Environment, Food and Rural Affairs in England, the National Assembly for Wales, and the Scottish Executive in Scotland.
b) Countryside bodies
England-Natural England, Environment Agency. Wales-Countryside Council for Wales, Environment Agency. Scotland-Scottish Natural Heritage, Scottish Environmental Protection Agency. And any other body designated by statutory provision as having specific environmental responsibilities.
c) Determination-Taken from Regulation 15 "Determi-nation of applications" and is the process by which we make our decision about the application for consent.
d) Forestry projects-the project work that the Forestry Commission must assess under these Regulations are deforestation (conversion to another land use), afforestation, forestry roads and forestry quarries.
e) Opinion-Our consideration of the proposals from which we will decide whether or not the project is a relevant one. If it is, the applicant must apply for consent and provide an ES.
f) Relevant project-A forestry project (afforestation, deforestation, forest roads works and forest quarry works) that is likely, by virtue of factors such as its nature, size and location, to have a significant effect on the environment and as such requires the FC's consent.
g) Scoping-A gathering of all consultees and other interested parties to discuss and agree the significant issues of concern that require to be addressed by an applicant when preparing an Environmental Statement.
h) Screening-the process by which the Forestry Commission decides whether a project "is likely to have significant effects on the environment by virtue, inter alia, of its nature, size or location". This process is a distinct one from scoping.
i) Publicity-The Forestry Commission maintains a web-based EIA Register that gives details of all the decisions we make under these Regulations.

Applicants are also required to advertise details of any application for consent.

j) Thresholds-Area limits set by the Regulations below which it is not expected that the project will have a significant effect on the environment.

The Forestry Commission will assess whether:

o the proposed project is one of the following categories-afforestation, deforestation, forest roads or quarries;

o the area is above the relevant threshold (includes extensions to similar areas of work);

o the project is likely to have a significant effect on the environment.

Our consent will be required to proceed with the work if your proposals meet each of these requirements. You will be asked to provide an Environmental Statement to allow us to decide whether to give consent to the project. Under the 1999 Regulations, proposals are considered to be relevant projects (i.e. to require an EIA) if they fall within the categories listed below and the work proposed is likely to have a significant effect on the environment.

a) Initial afforestation: creating new woods and forests by planting trees (on an area that has not had trees for many years). This category includes using direct seeding or natural regeneration, planting Christmas trees and short rotation coppice;
b) Deforestation: conversion of woodland to another type of land use (e.g.heathland);
c) Forest roads: constructing forestry roads, including those within a forest and those leading to a forest;
d) Forest quarries: quarrying to obtain material (rock, sand and gravel) for the formation, alteration or maintenance of forest roads.

We may serve you with an Enforcement Notice if you carry out work on a project that would have required our consent. This notice will require you to comply with the Regulations. Situations may arise where one of the above forestry projects forms part of a wider development that requires Planning Permission.

In these circumstances, any necessary EIA will usually not be dealt with under the Forestry EIA Regulations but under the parallel Town and Country Planning EIA Regulations.

Where a project has been completed since 6 September 1999, work proposals of the same type (including that on land in different ownerships) that extend it beyond the thresholds may need our consent. This also applies where the accumulated area of all adjacent existing projects plus the new proposals, exceeds the thresholds.

PUBLIC PARTICIPATION IN ENVIRONMENTAL DECISION MAKING

The title of this study, poses several questions: What is the 'public'? What is 'participation'? What are the 'decisions' to which it refers? Starting with the last and working backwards, EDM refers to any process of decision-making where consequent significant environmental impacts are a possibility. This includes law making, planning, strategic planning, resource management planning, licensing of industry e.g. IPPC, environmental assessment (EIA), spatial planning etc. EDM can be even more complex than decision-making on other public issues.

First, environmental impacts do not respect property, jurisdiction or boundaries.

Second, EDM can involve government agencies as both manager and regulator.

Thirdly, environmental issues can provide especially heated value conflicts that require value trade offs.

The level at which the public is involved varies with the relevant legislation, and the attitude of the other stakeholders. Often it just means informing the public of a previously made decision and asking for comments, which may or may not be heeded. Sometimes it means informed consultation. Here there is exchange of information prior to the relevant authority's reasoned decision making, and all inputs are included and seen to be included. An example of this being the River Basin Management process under Article 14 of the Water Framework Directive as described by Judith Cuff (Cuff, J. 2001). Another example would be the Bantry Bay Coastal Zone Charter. Public participation, at its apex, could also mean that the public itself, in consultation with the relevant bodies, makes the final decision itself. Examples of this can be found under the auspices of Local Agenda 21. For public participation to be effective at any level, it requires the public to be well informed and kept aware of the possibility of participation. This requires a pro-active approach from industry and the relevant public bodies.

What then is 'the public'? The public is often treated as a unitary body, whereas in reality it is a collection of numerous continually shifting interests and alliances (Ortolano), which may be in conflict with each other. The term is used as a "catch-all to describe those with an interest in a decision, other than a proponent, operator, or responsible authority". (Petts and Leach). The individuals making up a public may be involved as individuals or as members of organisations. They may become involved due to their proximity, economics, social or environmental issues, values, etc. By contrast, stakeholders, of which the public is one, are literally those with a stake in an issue and may include non-governmental organizations (NGO's), government or its agents, industry, individuals, communities etc.

Stakeholders do not always want to be involved in an EDM process, but they have the right to know if their interests are affected. They may want to become involved at a different stake of the EDM process.

STATUS AND TRENDS OF DROUGHT AND DESERTIFICATION

Two thirds of Africa is classified as deserts or drylands. These are concentrated in the Sahelian region, the Horn of Africa and the Kalahari in the south. Africa is especially susceptible to land degradation and bears the greatest impact of drought and desertification. It is estimated that two-thirds of African land is already degraded to some degree and land degradation affects at least 485 million people or sixty-five percent of the entire African population. Desertification especially around the Sahara has been pointed out as one the potent symbols in Africa of the global environment crisis. Climate change is set to increase the area susceptible to drought, land degradation and desertification in the region.

Under a range of climate scenarios, it is projected that there will be an increase of 5-8 per cent of arid and semi Arid lands in Africa. Estimates from individual countries report increasing areas affected by or prone to desertification. It is estimated that 35 percent of the land area (about 83,489 km2 or 49 out of the 138 districts) of Ghana is prone to desertification, with the Upper East Region and the eastern part of the Northern Region facing the greatest hazards. Indeed a recent assessment indicates that the land area prone to desertification in the country has almost doubled during recent times. Desertification is said to be creeping at an estimated 20,000 hectares per year, with the attendant destruction of farmlands and livelihoods in the country.

Seventy percent of Ethiopia is reported to be prone to desertification, 26 while in Kenya, around 80 percent of the land surface is threatened by desertification. Estimates of the extent of land degradation within Swaziland suggest that between 49 and 78 per cent of the land is at risk, depending on the assessment methodology used (Government of Swaziland, 2000). Nigeria is reported to be losing 1,355 square miles (1mile =1.6km) of rangeland and cropland to desertification each year. This affects each of the 10 northern states of Nigeria. It is estimated that more than 30 per cent of the land area of Burundi, Rwanda, Burkina Faso, Lesotho and South Africa is severely or very severely degraded. These rates and extent of land degradation/desertification undermine and pose serious threats to livelihoods of millions of people struggling to edge out of poverty. They also cripple provision of land resources-based ecosystem services that are vital for a number of development sectors.

Drought is one of the most important climate-related disasters in Africa. Climate change is set to exacerbate occurrence of climate related disasters including drought. A study from Bristol University projects that areas of western Africa were at most risk from dwindling freshwater supplies and droughts as a result of rising temperatures.30 Current climate scenarios predict that the driest regions of the world will become even drier, 31 signalling a risk of persistence of drought in many parts of Africa (arid, semi-arid and dry sub humid areas) which will therefore bear greater and sustained negative impacts.

Impact of Drought and Desertification

It is common knowledge that land degradation and desertification constitutes major causes of 25. forced human migration and environmental refugees, deadly conflicts over the use of dwindling natural resources, food insecurity and starvation, destruction of critical habitats and loss of biological diversity, socio-economic instability and poverty and climatic variability through reduced carbon sequestration potential. The impacts of drought and desertification are among the most costly events and processes in Africa. The widespread poverty, the fact that a large share of Africa's economies depend on climate-sensitive sectors mainly rain fed agriculture, poor infrastructure, heavy disease burdens, high dependence on and unsustainable exploitation of natural resources, and conflicts render the continent especially vulnerable to impacts of drought and desertification.

The consequences are mostly borne by the poorest people and the Small Island Developing States (SIDS). In the region, women and children in particular, bear the greatest burden when land resources are degraded and when drought sets in. As result of the frequent droughts and desertification, Africa has continued to witness food insecurity including devastating famines, water scarcity, poor health, economic hardship and social and political unrest.32 The gravity of drought and desertification impacts in the region is demonstrated by the following examples.

CAUSES OF DROUGHT AND DESERTIFICATION

The underlying cause of most droughts can be related to changing weather patterns manifested through the excessive build up of heat on the earth's surface, meteorological changes which result in a reduction of rainfall, and reduced cloud cover, all of which results in greater evaporation rates. The resultant effects of drought are exacerbated by human activities such as deforestation, overgrasing and poor cropping methods, which reduce water retention of the soil, and improper soil conservation techniques, which lead to soil degradation. Desertification is caused by multiple direct and indirect factors. It occurs because drylands ecosystems are extremely vulnerable to over-exploitation and inappropriate land use that result in underdevelopment of economies and in entranced poverty among the affected population.

Whereas over cultivation, inappropriate agricultural practices, overgrasing and deforestation have been previously identified as the major causes of land degradation and desertification, it is in fact a result of much deeper underlying forces of socio-economic nature, such as poverty and total dependency on natural resources for survival by the poor. It is also true to reiterate that desertification problems are best understood within the dictates of disparities of income and access to or ownership of resources. Consequently, the causes of desertification are more complex to unravel. Desertification is driven by a group of core variables, most prominently climatic factors (Yang and Prince 2000; Hulme and Kelly 1993) that lead to reduced rainfall (Rowell et al. 1992) and human activities involving technological factors, institutional and policy factors, and economic factors (UNCCD 2004) in addition to population pressures, and land use patterns and practices. The technological factors include innovations such as the adoption of water pumps, boreholes, and dams.

The institutional and policy factors include agricultural growth policies such as land distribution and redistribution (AIBS 2004). These variables drive proximate causes of desertification such as the expansion of cropland and overgrasing, the extension of infrastructure, increased aridity, and wood extraction.Since most economies of African countries are mostly agro-based, a greater proportion of the desertification problems in rural areas are a result of poverty related agricultural practices and other land use systems. Inappropriate farming systems such as continuous cultivation without adding any supplements, overgrasing, poor land management practices, lack of soil and water conservation structures, and high incidence of indiscriminate bushfires lead to land degradation and aggravate the process of desertification.

These factors prevail in many parts of the region. In Uganda, as a result of overgrasing in its drylands known as the "cattle corridor," soil compaction, erosion and the emergence of low-value grass species and vegetation have subdued the land's productive capacity, leading to desertification. In the Gambia, it is reported that fallow periods have been reduced to zero on most arable lands. Between 1950 and 2006, the Nigerian livestock population grew from 6 million to 66 million, a 11-fold increase. The forage needs of livestock exceed the carrying capacity of its grasslands. It is reported that overgrasing and over-cultivating are converting 351,000 hectares of land into desert each year. The rates of land degradation are particularly acute when such farming practices are extended into agriculture on marginal lands such as arid and semi rid lands, hilly and mountainous areas and wetlands. Deforestation, especially to meet energy needs and expand agricultural land is another serious direct cause of desertification in the region.

Globally, there is evidence demonstrating a heavy negative impact of the energy sector on forest and other vegetation cover and land productivity. More than 15 million hectares of tropical forests are depleted or burnt every year in order to provide for small-scale agriculture or cattle ranching, or for use as fuel wood for heating and cooking. Biomass constitutes 30 percent of the energy used in Africa and over 80 percent used in many sub-Saharan countries such as Burundi (91 percent), Rwanda and Central Africa Republic (90 percent), Mozambique (89 percent), Burkina Faso (87 percent), Benin (86 percent), Madagascar and Niger (85 percent).8 Production and consumption of fuel wood is said to have doubled in the last 30 years of the 20th century and is rising by 0.5 percent every year. This high dependence on biomass fuel has resulted into an alarming rate of tree felling and deforestation, which is exposing large tracts of land to desertification. In Ghana, where the population density has reached 77 persons per km^2, 70 percent of the firewood and charcoal needed for domestic purposes comes from the savannah zones, as a result destroying 20,000 ha of woodland per annum. In Uganda where 90 percent of the population lives in rural areas and directly depends on land for cultivation and grasing, forestland shrank from 45 percent of the country's surface area to 21 percent between 1890 and 200011. In Nigeria where more than 70 per cent of the nation's population depends on fuel wood, it is feared that the country might be left with no forest by 2010 owing to the present level of deforestation activities. Already it is estimated that more than 13 million tonnes of soil are washed away into the sea annually. It is also feared that if the current rate of tropical forests deforestation is maintained, the tropical forests could be almost entirely harvested by the year 2050, thus devastatingly contributing to climate change, loss of biodiversity, land degradation and desertification. The above direct causes of desertification are driven by a complex set of underlying factors including the high levels of poverty in the region, high population growth rates, poor natural resources tenure and access regimes, conflicts, and climate change. Without alternatives poor people are forced to exploit land resources including fragile lands, for survival (food production, medicine, fuel, fodder, building materials and household items).

Given that most drylands in Africa are poverty hotspots as well, the risk of desertification is high in many of these areas, as the poor inevitably become both the victims and willing agents of environmental damage and desertification. In Sub-Saharan Africa alone 270 million people live in absolute poverty. In Uganda, over 40 percent of the pastoralists who constitute the majority in the country's drylands, live below the poverty line. High population growth increases pressure on limited and fragile land resources.

The rural population living in drylands in Africa is estimated to be 325 million. This breeds favourable conditions for deforestation and overexploitation of land that lead to land degradation as a large and growing rural population, struggling to survive in a limited natural resource base result in the over-utilisation of the available natural resources. For instance the Nigeria's human population which grew from 33 million in 1950 to 134 million in 2006, a fourfold expansion has forced farmers to plough marginal land under the pressure to meet food needs. As a result of this, the country is slowly turning into a desert. According to the New York Times, Niger's population has doubled in the last 20 years. Each woman bears about seven children, giving the country one of the highest growth rates in the world. Given that 90 percent of Niger's people live off agriculture, this population is exerting great pressure on the less than 12 percent of its land that can be cultivated. Insecure and unclear land and other natural resources tenure and access rights are some of the main reasons the natural resources end-users are unwilling to invest in long-term sustainable land management (SLM). For instance it is reported that in Uganda, insecurity of land tenure in parts of the cattle corridor under mailo and communal land ownership systems does not encourage farmers to invest in sustainable land management practices.

MARINE COASTAL ECOSYSTEMS

Seashores throughout the world are subject to increasing pressures from residential, recreational, and commercial development. These stresses may become more severe, for human population in the vicinity of sea-coasts is growing at twice the inland rate. Some of the pressures that we exert on coastal ecosystems are summarized in the accompanying box. All can increase the growth of algae. Among the possible consequences of disruption in almost any marine ecosystem is an increase in the opportunistic pathogens that can abet the spread of human disease, sometimes to widespread proportions. One example is cholera.

Cholera

We often think of our modern world as cleansed of the epidemic scourges of ages past. But cholera —an acute and sometimes fatal disease that is accompanied by severe diarrhea— affects more nations today than ever before. The Seventh Pandemic began when the El Tor strain left its traditional home in the Bay of Bengal in the 1960s, travelled to the east and west across Asia, and in the 1970s penetrated the continent of Africa. In 1991, the cholera pandemic reached the Americas, and during the first eighteen months more than half a million cases were reported in Latin America, with 5,000 deaths.

Rapid institution of oral rehydration treatment—with clean water, sugar, and salts—limited fatalities in the Americas to about one in a hundred cases. The epidemics also had serious economic consequences. In 1991, Peru lost $770 million in seafood exports and another $250 million in lost tourist revenues because of the disease.

The microbe that transmits cholera, Vibrio cholerae, is found in a dormant or "hibernating" state in algae and microscopic animal plankton, where it can be identified using modern microbiological techniques. But once introduced to people—by drinking contaminated water or eating contaminated fish or shellfish— cholera can recycle through a population, when sewage is allowed to mix with the clean water supply.

Five years ago, in late 1992, a new strain of Vibrio cholerae—O139 Bengal—emerged in India along the coast of the Bay of Bengal. With populations unprotected by prior immunities, this hardy strain quickly spread through adjoining nations, threatening to become the agent of the world's Eighth Cholera Pandemic. For a time, in 1994, El Tor regained dominance. But by 1996, O139 Bengal had reasserted itself. The emergence of this new disease, like all others, involved the interplay of microbial, human host, and environmental factors.

The largest and most intense outbreak of cholera ever recorded occurred in Rwanda in 1994, killing over 40,000 people in the space of weeks, in a nation already ravaged by civil war and ethnic strife. The tragedy of cholera in Rwanda is a reminder of the impacts of conflict and political instability on public health and biological security—just as epidemics may, in turn, contribute to political and economic stability.

Is The Ocean Warming?

Surface temperatures of the ocean have warmed this century, and a gradual warming of the deep ocean has been found in recent years in oceanographic surveys carried out in the tropical Pacific, Atlantic, and Indian Oceans, and at both poles of the Earth. These findings could be indicative of a long-term trend. Corresponding temperature measurements of the sub-surface earth, in cores drilled deep into the Arctic tundra, show a similar effect.

The water that evaporates from warmer seas, and from vegetation and soils of a warmer land surface, intensifies the rate at which water cycles from ocean to clouds and back again. In so doing it increases humidity and reinforces the greenhouse effect. Warm seas are the engines that drive tropical storms and fuel the intensity of hurricanes. More high clouds can also contribute to warmer nights by trapping out-going radiation.

Some Biological Impacts

A warmer ocean can also harm marine plankton, and thus affect more advanced forms of life in the sea. A northward shift in marine flora and fauna along the California coast that has been underway since the 1930s has been associated with the long-term warming of the ocean over that span of time.

Warming—when sufficient nutrients are present—may also be contributing to the proliferation of coastal algal blooms. Harmful algal blooms of increasing extent, duration, and intensity—and involving novel, toxic species—have been reported around the world since the 1970s. Indeed, some scientists feel that the worldwide increase in coastal algal blooms may be one of the first biological signs of global environmental change.

Warm years may result in a confluence of adverse events. The 1987 El Nino was associated with the spread and new growth of tropical and temperate species of algae in higher northern and southern latitudes. Many were toxic algal blooms. In 1987, following a shoreward intrusion of Gulf Stream eddies, the dinoflagellate Gymnodimuim breve, previously found only as far north as the Gulf of Mexico, bloomed about 700 miles north, off Cape Hatteras, North Carolina, where it has since persisted, albeit at low levels. Forty-eight cases of neurological shellfish poisoning occurred in 1987, resulting in an estimated $25 million loss to the seafood industry and the local community. In the same year, anomalous rain patterns and warm Gulf Stream eddies swept unusually close to Prince Edward Island in the Gulf of St. Lawrence. The result, combined with the run-off of local pollutants after heavy rains, was a bloom of toxic diatoms. For the first time, domoic acid was produced from these algae, and then ingested by marine life. Consumption of contaminated mussels resulted in 107 instances of amnesic shellfish poisoning, from domoic acid, including three deaths and permanent, short-term memory loss in several victims.

Also in 1987, there were major losses of sea urchin and coral communities in the Caribbean, a massive sea grass die-off near the Florida Keys, and on the beaches of the North Atlantic coast, the death of numerous dolphins and other sea mammals. It has been proposed that the combination of algal toxins, chlorinated hydrocarbons like PCBs, and warming may have lowered the immunity of organisms and altered the food supply for various forms of sea life, allowing *morbilli* (measles-like) viruses to take hold.

The 1990s

For five years and eight months, from 1990 to 1995, the Pacific Ocean persisted in the warm El Nino phase, which was most unusual, for since 1877 none of these distinctive warmings had lasted more than three years. Both anomalous phases—with either warmer (El Nino) or colder (La Nina) surface waters— bring climate extremes to many regions across the globe. With the ensuing cold (La Nina) phase of 1995-1996, many regions of the world that had lived with drought during the El Nino years were now besieged with intense rains and flooding. Just as in Colombia, flooding in southern Africa was accompanied by an upsurge of vector-borne diseases, including malaria. Other areas experienced a climatic switch of the opposite kind, with drought and wildfires replacing floods. During 1996 world grain stores fell to their lowest level since the 1930s. Weather always varies; but increased variability and rapid temperature fluctuations may be a chief characteristic of our changing climate system. And increased variability and weather volatility can have significant consequences for health and for society.

Decadal Variability

The cumulative meteorological and ecological impacts of the prolonged El Nino of the early 1990s have yet to be fully evaluated, and another is now upon us. In 1995, warming in the Caribbean produced coral bleaching for the first time in Belize, as sea surface temperatures surpassed the 29°C (84°F) threshold that may damage the animal and plant tissues that make up a coral reef. In 1997, Caribbean sea surface temperatures reached 34°C (93°F) off southern Belize, and coral bleaching was accompanied by large mortalities in starfish and other sea life. Coral diseases are now sweeping through the Caribbean, and diseases that perturb marine habitat, such as coral or sea grasses, can also affect the fish stocks for which these areas serve as nurseries. A pattern of greater weather variability has begun and is expected to persist with the El Nino of 1997 and 1998. Since 1976, such anomalies in Pacific Ocean temperatures and in weather extreme events have become more frequent, more intense, and longer lasting than in the preceding 100 years, as indicated in records kept since 1877.

DISCONTINUITIES AND INSTABILIBY

The common perception that the natural world changes only gradually can be misleading, for discontinuities abound. Animals switch abruptly between two states—awake and asleep—that are sharply divided and marked by qualitative differences in levels of activity in the central nervous system. Water can rapidly change from vapour to liquid to solid. Ecosystems have equilibrium states that are also at times abruptly interrupted. An extensive fire in an old growth forest, for example, can radically change the types of plants and animals within it.

Climate regimes can also change surprisingly fast. Recent analyses of Greenland ice cores indicate that significant shifts, called rapid climate change events (RCCEs), have taken place in the past in the span of but several years—

not centuries, as was previously believed. While the oceans may serve as a buffer against sudden climate change, this mechanism may be limited, for some of the RCCEs seem to be associated with abrupt changes in ocean circulation.

The climate system exhibits equilibrium states as well, of which three may have been most common: when the poles of the Earth were covered with small, medium, or large ice caps. The present, Holocene period of the last 10,000 years—with medium-size caps and an average global temperature of 15°C (about 60°F)— has been associated with the development of modern agriculture and advancing civilization.

But our present climate regime may be becoming less stable. Increased variance—that is, more extreme swings—in natural systems is inversely related to how stable and balanced the systems are, and how sensitive they are to perturbations. Wider and wider variations can occur as a system moves away from its equilibrium state.

Trends in the 20th Century

The gradual warming that characterized the climate during the first four decades of the present century, for example, was accompanied by substantial temperature variability, as borne out in the record of degree-heating-days in the U.S. grain belt. The ensuing cooling trend from 1940 to the mid 1970s showed less variability. From 1976 to the present day, the variability—apparent in hot and cold spells, drought, and floods—has again increased. Greenland ice core records suggest that the last time the Earth warmed abruptly, ending the last Ice Age, there was also a pattern of increased variability.

The connection between human health and environmental stability increases our need for a better understanding of the present state of the global climate system. There are several unanswered questions regarding the system's stability. Was the drift toward earlier springtimes that began in this country in the 1940s indicative of the first minor readjustment in the climate regime? Are the more frequent and intense El Nino events since the mid 1970s another such indicator? Has the baseline of ocean temperatures shifted? Does the present climatic volatility—evident in altered weather and precipitation patterns—increase the potential for an abrupt "jump" in the climate system? And might further stresses lead to abrupt discontinuities of the type found in the Greenland ice cores, when the last Ice Age rapidly came to an end?

Bibliography

Adrian E. : *Principles of Geodynamics*, Berlin, Springer-Verlag, 1958.

Ager, D. V., *The Nature of the Stratigraphic Record*, London, Macmillan Press, 1981.

Albritton, C. C. : *The Fabric of Geology: Reading*, Mass., Addison-Wesley Publishing Co., 1963.

Allègre C.J.: *Isotope Geology*, Cambridge University Press, 2008.

Alok K. : *Petrology and Genesis of Igneous Rocks*, Narosa, Delhi, 2007.

Arnold, C. A. : *Introduction to Paleobotany*, New York, McGraw-Hill, 1947.

Babar Md. : *Hydrogeomorphology : Fundamentals Applications and Techniques*, New India Pub, Delhi, 2005.

Ballabh. Parikshit : *An Introduction to Geology*, Cyber Tech Pub, Delhi, 2009.

Baruah Akhil : *Advanced Morphology of Angiosperms*, Aavishkar, Delhi, 2008.

Benn, Douglas I. and David J. A. Evans: *Glaciers and Glaciation*. London; Arnold, 1998.

Bhagwat, S.B. : *Foundation of Geology*, Global Vision Pub, Delhi, 2009.

Bowles, J.: *Foundation Analysis and Design*, McGraw-Hill Publishing Company, 1981.

Bucher, Kurt: *Petrogenesis of Metamorphic Rock*, Springer, 2002.

Carey, S. Warren : *Theories of the Earth and Universe*, Stanford, Stanford Univ. Press, 1988.

Dickin A.P.: *Radiogenic Isotope Geology*, Cambridge University Press, 2005.

Drever, James: *The Geochemistry of Natural Waters*. New Jersey: Prentice Hall, 2002.

Duckworth W L H : *Morphology and Anthropology : A Handbook for Students*, Cosmo, Delhi, 2006.

Eskola P.: *The Mineral Facies of Rocks*, Norsk. Geol. Tidsskr, 1920.

Faure G., Mensing T.M.: *Isotopes: Principles and Applications*, John Wiley & Sons, 2004.

Gillen, Cornerlius: *Metamorphic Geology : an Introduction to Tectonic and Metamorphic Processes*, London; Boston: G. Allen & Unwin, 1982.

Gillespie, C. C.: Genesis and Geology: New York, Harper, 1951.

Glenn T. Trewartha : *Elements of Geography*, New York, McGraw Hill, 1941.

Greve, Ralf and Heinz Blatter: *Dynamics of Ice Sheets and Glaciers*. Berlin etc.; Springer, 2009.

Gurjar R.D. : *Geomorphology and Environmental Sustainability*, Concept, Delhi, 2005.

Gurugnanam, B. : *Essentials of Hydrogeology*, New India Pub, Delhi, 2009.

Hallam, A.: *A Revolution in the Earth Sciences*, New York, Oxford University Press, 1973.

Hambrey, Michael and Jürg Alean: *Glaciers*. Cambridge and New York; Cambridge University Press, 2004.

Hoefs J.: *Stable Isotope Geochemistry,* Springer Verlag, 2004.

Holtz, R. and Kovacs, W.: *An Introduction to Geotechnical Engineering,* Prentice-Hall, 1981.

Hooke, Roger : *Principles of Glacier Mechanics*. Cambridge and New York; Cambridge University Press, 2005.

Kramer, Steven L.: *Geotechnical Earthquake Engineering,* Prentice-Hall, 1996.

Lyell, C. : *Principles of Geology,* Boston, Little, Brown, 1853.

Mallik, T K : *Marine Geology : A Scenario Around Indian Coasts,* New Academic Pub, Delhi, 2008.

Marshak, Stephen: *Essentials of Geology,* W. W. Norton, 2009.

Mathur, S M : *Concise Glossary of Geology,* Scientific, Delhi, 2007.

Mitchell, James K. & Soga, K.: *Fundamentals of Soil Behavior,* John Wiley & Sons, 2005.

Norton, W.H. : *A Textbook of Geology : Elements and Theories,* Dominant, Delhi, 2011.

Prasad, C.V.R.K. : *Elementary Exercises in Geology,* Universities Press, Delhi, 2005.

Pullaiah T. : *Taxonomy of Angiosperms,* Regency Pub, Delhi, 2007.

Rollinson, H.R.: *Using Geochemical Data: Evaluation, Presentation, Interpretation* Longman Scientific & Technical. 1993.

Rushton, K.R.: *Groundwater Hydrology: Conceptual and Computational Models*. John Wiley and Sons Ltd. 2003.

Sawant, P. T. : *Engineering and General Geology,* New India Publishing Agency, Delhi, 2011.

Sharma, V K : *Origin and Development of Geology,* Vista International Pub, Delhi, 2008.

Sharp Z.: *Principles of Stable Isotope Geochemistry,* Prentice Hall, 2006.

Sinha Rama Kant : *Practical Taxonomy of Angiosperms,* I.K. International, Delhi, 2010.

Vernon, Ron H.: *A Practical Guide to Rock Microstructure,* Oxford University Press, Oxford, 1996.

Index

O

P

Q

R

S

T

❑❑❑